AF614707

Methods in Molecular Biology

Series Editor
John M. Walker
School of Life Sciences
University of Hertfordshire
Hatfield, Hertfordshire, AL10 9AB, UK

For further volumes:
http://www.springer.com/series/7651

Innate DNA and RNA Recognition

Methods and Protocols

Edited by

Hans-Joachim Anders

Medizinische Klinik und Poliklinik IV, Ludwig Maximilians Universität München, München, Germany

Adriana Migliorini

Helmholtz Zentrum München German Research Center for Environmental Health, Institute for Diabetes and Regeneration Research (IDR), Munich, Germany

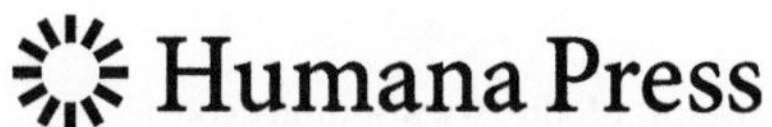

Editors
Hans-Joachim Anders
Medizinische Klinik und Poliklinik IV
Ludwig Maximilians Universität München
München, Germany

Adriana Migliorini
Helmholtz Zentrum München German Research Center for Environmental Health
Institute for Diabetes and Regeneration Research (IDR)
Munich, Germany

ISSN 1064-3745 ISSN 1940-6029 (electronic)
ISBN 978-1-4939-0881-3 ISBN 978-1-4939-0882-0 (eBook)
DOI 10.1007/978-1-4939-0882-0
Springer New York Heidelberg Dordrecht London

Library of Congress Control Number: 2014941619

Printed on acid-free paper

Humana Press is a brand of Springer
Springer is part of Springer Science+Business Media (www.springer.com)

Preface

The last decade has completely changed our understanding of pathogen recognition in terms of how the innate immune system manages to integrate molecular patterns of completely different pathogens into rather uniform innate immune responses. This immunostimulatory effect of pathogen-derived nucleic acids is of great importance for host defense as it not only initiates an immediate innate immune response but also serves as an adjuvant for the priming of adaptive immune responses that rapidly combat the pathogen upon subsequent exposures. This mechanism renders nucleic acid-based adjuvants interesting compounds for vaccination strategies. Although the structures of nucleic acids are strongly preserved from prokaryotes to mammals, the mammalian immune system has found ways to distinguish own nucleic acids from those of bacteria and viruses. This is necessary for immune tolerance and it is supported by compartmentalization of nucleic acids away from nucleic acid sensors, by epigenetic DNA and RNA modifications, by rapid opsonization of extracellular nuclear particles and their phagocytic clearance by macrophages. However, unfortunate combinations of genetic variants or certain drugs can compromise these protective mechanisms. This can turn tolerogenic autoantigen presentation into aberrant activation and proliferation of auto-reactive lymphocytes, with ensuing production of antinuclear antibodies and induction of autoimmune diseases like systemic lupus erythematosus and scleroderma.

The biological significance of nucleic acid immune recognition is not limited to host defense, vaccination, and autoimmunity. The diagnostic and therapeutic use of gene targeting nucleic acids, e.g., siRNAs, miRNAs, or aptamers, has to consider the immunostimulatory potential of certain nucleic acid formats. This implies either to avoid immune stimulation in therapeutic gene regulation with siRNA or aptamers, or to utilize immune stimulation to overcome tumor-associated immunosuppression, e.g., by the use of bifunctional siRNAs that combine a knock-down sequence with immunostimulatory structural features in one molecule.

This edition of Methods in Molecular Biology presents validated experimental strategies to dissect nucleic acid sensing in vitro and in vivo. It is meant as a resource for immunologists, molecular biologists, virologists, microbiologists, and any researcher that wants to know how the innate immune system handles nucleic acids from endogenous or foreign sources.

Several of the authors have collaborated within the Research Training Group 1202 "Oligonucleotides in cell biology and therapy" that has been funded by the Deutsche Forschungsgemeinschaft from 2005 to 2014 at the Ludwig-Maximilians-Universität in Munich.

Munich, Germany

Adriana Migliorini
Hans-Joachim Anders

Contents

Part II Analysis of Nucleic Acid Sensing In-Vivo

Contributors

Hans-Joachim Anders • *Medizinische Klinik und Poliklinik IV, Ludwig Maximilians Universität München, München, Germany*

David Anz • *Division of Clinical Pharmacology, Department of Internal Medicine, Ludwig-Maximilian University of Munich, Munich, Germany*

Sabine Bartel • *Comprehensive Pneumology Center, Helmholtz Zentrum München, Ludwig-Maximilians Universität and Asklepios Clinic Gauting, Munich, Germany*

Stefan Bauer • *Institut für Immunologie, BMFZ, Philipps-Universität Marburg, Marburg, Germany*

Isabelle Bekeredjian-Ding • *Institute for Medical Microbiology, Immunology and Parasitology (IMMIP), University Hospital Bonn, Bonn, Germany*

Robert Besch • *Department of Dermatology and Allergology, Ludwig-Maximilian University, Munich, Germany*

Ramon G.B. Bonegio • *Renal Section, Department of Medicine, Boston University School of Medicine, Boston, MA, USA*

Filiz Civril • *Department of Biochemistry at the Gene Center, Ludwig-Maximilians-University Munich, Munich, Germany; Graduate School for Quantivative Biosciences Munich, Ludwig-Maximilians-University Munich, Munich, Germany*

Stefan Dehmel • *Comprehensive Pneumology Center, Helmholtz Zentrum München, Ludwig-Maximilians Universität and Asklepios Clinic Gauting, Munich, Germany*

Rika Draenert • *Department of Infectious Diseases, Medizinische Klinik und Poliklinik IV, Klinikum der Universität München, Munich, Germany*

Katharina Eisenächer • *II. Medical Department, Klinikum rechts der Isar, Technical University Munich, Munich, Germany*

Marion Goldeck • *Institute for Clinical Chemistry and Clinical Pharmacology, University Hospital, University of Bonn, Bonn, Germany*

Jing Gong • *Cancer Research Center, School of Medicine, Shandong University, Jinan, China*

Gunther Hartmann • *Institute for Clinical Chemistry and Clinical Pharmacology, University Hospital, University of Bonn, Bonn, Germany*

Alexander Heiseke • *II. Medical Department, Klinikum rechts der Isar, Technical University Munich, Munich, Germany*

Karl-Peter Hopfner • *Department of Biochemistry at the Gene Center, Ludwig-Maximilians-University Munich, Munich, Germany; Graduate School for Quantivative Biosciences Munich, Ludwig-Maximilians-University Munich, Munich, Germany*

John M. Hoppe • *Nephrologisches Zentrum, Medizinische Klinik und Poliklinik IV, Klinikum der Universität München, Ludwig-Maximilians-University Munich, Munich, Germany*

Veit Hornung • *Institute of Molecular Medicine, University Hospital, University of Bonn, Bonn, Germany*

Theresa Kaeuferle • *Dr. von Hauner'sche Kinderklinik, Klinikum der Universität München, Munich, Germany*

ANDREAS KAUFMANN • *Institut für Immunologie, BMFZ, Philipps-Universität Marburg, Marburg, Germany*

VIKTOR H. KOELZER • *Clinical Pathology Division and Translational Research Unit, Institute of Pathology, University of Bern, Bern, Switzerland*

SUSANNE KRAUSS-ETSCHMANN • *Comprehensive Pneumology Center, Helmholtz Zentrum München, Ludwig-Maximilians Universität and Asklepios Clinic Gauting, Munich, Germany*

ANNE KRUG • *II. Medical Department, Klinikum rechts der Isar, Technical University Munich, Munich, Germany*

MACIEJ LECH • *Nephrologisches Zentrum, Klinische Biochemie, Medizinische Klinik und Poliklinik IV, Klinikum der Universität München, Munich, Germany*

ANDREAS LINDER • *Division of Clinical Pharmacology, Ludwig-Maximilian University Munich, Munich, Germany*

NICOLAS LINDER • *Division of Clinical Pharmacology, Ludwig-Maximilian University Munich, Munich, Germany*

FANNY MATHEIS • *Department of Dermatology and Allergology, Ludwig-Maximilian University, Munich, Germany*

ADRIANA MIGLIORINI • *Helmholtz Zentrum München German Research Center for Environmental Health, Institute for Diabetes and Regeneration Research (IDR), Munich, Germany*

GERNOT NEES • *Institut für Immunologie, BMFZ, Philipps-Universität Marburg, Marburg, Germany*

MORITZ RAPP • *Division of Clinical Pharmacology, Medizinische Klinik und Poliklinik IV, Klinikum der Universität München, Munich, Germany*

ANDREA RIBEIRO • *Medizinische Klinik und Poliklinik Campus Innenstadt, Klinikum der LMU, Munich, Germany*

IAN R. RIFKIN • *Renal Section, Department of Medicine, Boston University School of Medicine, Boston, MA, USA*

JULIA ROIDER • *Department of Infectious Diseases, Medizinische Klinik und Poliklinik IV, Klinikum der Universität München, Munich, Germany*

SIMONE ROMOLI • *Nephrologisches Zentrum, Medizinische Klinik und Poliklinik IV, Klinikum der Universität München, Ludwig-Maximilians-University Munich, Munich, Germany*

SIMON ROTHENFUSSER • *Division of Clinical Pharmacology, Ludwig-Maximilian University Munich, Munich, Germany*

MARTIN SCHLEE • *Institute for Clinical Chemistry and Clinical Pharmacology, University Hospital, University of Bonn, Bonn, Germany*

ANDREAS SCHMIDT • *Division of Clinical Pharmacology, Ludwig-Maximilian University Munich, Munich, Germany*

MAX SCHNURR • *Division of Clinical Pharmacology, Medizinische Klinik und Poliklinik IV, Klinikum der Universität München, Munich, Germany*

VOLKER VIELHAUER • *Nephrologisches Zentrum, Medizinische Klinik und Poliklinik IV, Klinikum der Universität München, Ludwig-Maximilians-University Munich, Munich, Germany*

THOMAS VOLLBRECHT • *Department of Infectious Diseases, Medizinische Klinik und Poliklinik IV, Klinikum der Universität München, Munich, Germany*

AMANDA A. WATKINS • *Renal Section, Department of Medicine, Boston University School of Medicine, Boston, MA, USA*

TIANDI WEI • *State Key Laboratory of Microbial Technology, School of Life Sciences, Shandong University, Jinan, China*
MARKUS WÖRNLE • *Medizinische Klinik und Poliklinik Campus Innenstadt, Klinikum der LMU, Munich, Germany*
SASKIA ZIEGLER • *Department for Infectious Diseases, Medical Microbiology and Hygiene, University Hospital Heidelberg, Heidelberg, Germany*

Part I

Analysis of Viral Nucleic Acid Sensing In-Silico and In-Vitro

Chapter 1

Detection of RNA Modifications by HPLC Analysis and Competitive ELISA

Gernot Nees, Andreas Kaufmann, and Stefan Bauer

Abstract

Over 100 different RNA modifications exist that are introduced posttranscriptionally by enzymes at specific nucleotide positions. Ribosomal RNA (rRNA) and transfer RNA (tRNA) exhibit the most and diverse modifications that presumably optimize their structure and function. In contrast, oxidative damage can lead to random modifications in rRNA and messenger RNA (mRNA) that strongly impair functionality. RNA modifications have also been implicated in avoiding self-RNA recognition by the immune system or immune evasion by pathogens. Here, we describe the detection of RNA modifications by HPLC analysis and competitive ELISA.

Key words HPLC, Competitive ELISA, RNA, Nucleoside modification, Base methylation, 2′-O-Methylation, N1-Methylguanosine, 2′-O-Methylguanosine, 8-Hydroxyguanosine (8-OHG)

1 Introduction

Over 100 different RNA modifications have been identified and exist in all three kingdoms of life. A comprehensive listing of posttranscriptionally modified RNA nucleosides is found in the RNA Modification Database (http://mods.rna.albany.edu/) [1].

Especially ribosomal and transfer RNA (rRNA and tRNA) are abundantly modified. Examples are 2′-O-ribose methylation, base methylation, and the occurrence of pseudouridine. In the case of 2′-O-ribose methylation of rRNA, the methyltransferase (2′-O-MTase) fibrillarin utilizes small nucleolar RNAs (snoRNAs), so-called Box C/D snoRNAs that guide the enzyme complex to complementary regions of the rRNA for methylation [2]. Base methylation of rRNA and tRNA as well as 2′-O-ribose methylation of tRNA are carried out by position/sequence-specific methyltransferases independent of snoRNAs. For example, in *E. coli* uracil-5-methyltransferase (trmA) and guanine-7-methyltransferase (yggH/trmB) methylate specific bases in tRNA such as uridine 54 (m5U54) or guanosine 46 (m7G46), respectively [3–5]. In

Hans-Joachim Anders and Adriana Migliorini (eds.), *Innate DNA and RNA Recognition*, Methods in Molecular Biology, vol. 1169, DOI 10.1007/978-1-4939-0882-0_1, © Springer Science+Business Media New York 2014

contrast, the Gm18-2′-O methyltransferase (spoU/trmH) methylates the 2′-O-position of a conserved guanosine at position 18 of tRNA (Gm18) [6].

Interestingly, also messenger RNA (mRNA) is internally modified and carries N6-methyladenosine (m^6A) [7]. m^6A is present in mRNA of all higher eukaryotes tested, including mammals, plants, and insects. This modification occurs on average at 1–3 residues within a defined sequence context (e.g., GGACU) per typical mammalian mRNA molecule [7]. Recently it has been reported that m^6A sites are enriched near stop codons and in 3′ UTRs suggesting an important role in regulation of gene expression [8, 9].

Random RNA modifications may also occur by oxidation through reactive oxygen species (ROS) which are involved in killing of bacteria and cell signaling pathways [10–12]. Over 20 different purine and pyrimidine modifications formed by reactive oxygen species are known; however, 8-hydroxyguanosine (8-OHG) is the most prominent modification [13]. Of note, 8-hydroxyguanosine modification in mRNA leads to reduced protein levels and altered protein function due to ribosome stalling [14]. Interestingly, age-associated oxidative damage to RNA has been demonstrated in neurons and may play a role in neurodegeneration and other diseases [15].

Bacterial and viral RNA are potent stimulators of the innate immune system leading to immune activation [16]. RNA is recognized in the endosome by Toll-like receptors (TLR). TLR3 recognizes double-stranded viral RNA and mRNA, whereas TLR7 and TLR8 sense single-stranded RNA [17]. In contrast, cytoplasmic detection of viral RNA is mediated by the RNA helicases retinoic acid inducible gene-I (RIG-I) and melanoma-differentiation-associated gene 5 (MDA5) [18]. RIG-I recognizes 5′ triphosphate RNA [19, 20], whereas MDA5 is activated by higher-order RNA structures generated during viral infection [21].

The effect of ribonucleoside modifications on immunostimulation has been investigated only recently. RNA modifications such as 2′-O- or base methylation and the occurrence of pseudouridine negatively modify the immunostimulatory potential of in vitro transcribed RNA with respect to TLR3, TLR7, and TLR8 [22]. Also pseudouridine-containing or 2′-O-methylated synthetic RNA loses its immunostimulatory capacity via RIG-I [19]. We and others have further demonstrated that 2′-O-methylation of synthetic RNA and tRNA not only renders TLR7 ligands non-immunostimulatory but also converts modified RNA into a TLR7 antagonist [23–25].

Eukaryotic mRNA is not recognized by RIG-I or MDA5 due to a 5′ cap structure methylation at the N7 position of the capping guanosine residue (cap 0), and additional 2′-O-methylation(s) at the 5′-penultimate residue (cap 1) and sometimes also at adjoining residue (cap 2). Interestingly, some viruses that replicate in the cytoplasm (e.g., picornaviruses and coronaviruses) encode functions associated with the formation of a 5′ cap, which are homologous to those found in eukaryotic cells and also support immune evasion. Accordingly, 2′-O-methylation of viral mRNA cap structures by virus-encoded methyltransferases prevents recognition by MDA-5 or host restriction by interferon-induced proteins with tetratricopeptide repeats (IFIT) family members [26, 27].

In summary, the detection and characterization of modified ribonucleosides are important for understanding mechanisms of RNA-induced immune activation or immune evasion.

2 Materials

2.1 HPLC

1. HPLC system with a column heater and UV monitor. We use a Dionex UltiMate 3000 HPLC with autosampler (WPS-300SL), UV-Detector (VWD-3400RS), and column heater (TCC-300SD). Analysis is performed by the Chromeleon 6.80 SR10 Build 2818 software.
2. HPLC column used for separation is an analytical silica based octadecyl end-capped 25 cm×4.6 mm, 5 μm HPLC column. A silica based 2 cm octadecyl end-capped column served as a guard column.
3. Buffer A: 5 mM ammonium acetate, pH 6.0 (*see* **Note 1**).
4. Buffer B: 40 % (v/v) acetonitrile.
5. Buffer C: 66 % (v/v) methanol.
6. Buffer D: 40 % (v/v) methanol (*see* **Note 2**).

2.2 Standard Nucleosides

	Stock solution (mM)	Working solution (μM)
Cytidine	10	30
Uridine	10	30
2′-Deoxycytidine	10	70
Xanthosine	1	30
2′-O-Methylcytidine	10	30
Inosine	10	30
Guanosine	1	20
7-Methylguanosine	0.5	20

(continued)

	Stock solution (mM)	Working solution (μM)
8-Hydroxyguanosine	3	30
2′-O-Methyluridine	10	30
2′-Deoxyguanosine	1	15
Thymidine	10	200
2′-O-Methylguanosine	1	20
N_1-Methylguanosine	1	20
N_2-Methylguanosine	1	50
Adenosine	10	30
2′-Deoxyadenosine	10	80
2′-O-Methyladenosine	10	50
5′-Methyldeoxycytidine	5	50
5′-Methyluridine	10	50
N_6-Methyladenosine	1	10
N_6N_6-Dimethyladenosine	10	80

For solubilization *see* **Note 3**.

2.3 RNA Digestion

1. 10 mM Zinc chloride.
2. 300 mM Ammonium acetate, pH 5.3.
3. Tris Base: 100 mM, pH 8.3.
4. Magnesium acetate: 10 mM (*see* **Note 4**).
5. P1 endonuclease from *Penicillium citrinum*, SvP (Snake venom phosphodiesterase from *Crotalus adamanteus*, Sigma, cat. number P3134, 0.5 U/ml), and alkaline phosphatase (AP).
6. 50 μg RNA of interest (*see* **Note 5**).

2.4 Competitive ELISA for 8-Hydroxyguanosine (8-OHG)

1. Phosphate buffered saline (PBS).
2. Washing buffer: PBS + 0.05 % Tween20, store at RT.
3. Blocking buffer 1: washing buffer + 1 % bovine serum albumin fraction V (BSA), store at 4 °C.
4. Blocking buffer 2: blocking buffer 1 with addition of 2 % Sucrose and 0.05 % Casein hydrolysate, store at 4 °C.
5. Substrate buffer: 65 mM disodium hydrogen phosphate, 35 mM citric acid, pH 5.0, store at 4 °C.
6. Substrate: 20 ml substrate buffer + 20 mg O-Phenylenediamine dihydrochloride (OPD) + 20 μl 30 % H_2O_2. Store at 4 °C. Light sensitive, stable for 2 h.

7. Antibodies: Mouse Anti-8-OHG monoclonal antibody, Clone 15A3 (Cayman, cat. number 10011446, 0.65 mg/ml), store at −20 °C. Goat anti-mouse IgG, peroxidase-conjugated, (Jackson cat. number 115-035-062, 0.8 mg/ml), store at 4 °C.
8. Standard: 8-OHG (Cayman, cat. number 89300, 10 mg/ml, nucleoside concentration); Store at −20 °C.
9. BSA conjugated 8-hydrogyguanosine (OHG) (8.45 mg/ml, protein concentration) (*see* **Note 6**).
10. Photometer: Emax (Molecular Device).
11. Microtiter plates: MaxiSorp 96 well (Nunc).

3 Methods

3.1 Analytical HPLC

For detection of modified nucleotides in a given RNA sequence HPLC analysis of digested RNA is a suitable method with high sensitivity and specificity. Before starting the analysis, standard mixtures of nucleosides should be analyzed with the standard protocol to yield high resolution and reproducibility (*see* **Note 7**) (Fig. 1). The occurrence of RNA modifications within a RNA sample of interest (e.g., total RNA from a eukaryotic cell line) can be judged by comparison and overlay of the individual chromatograms (Fig. 1).

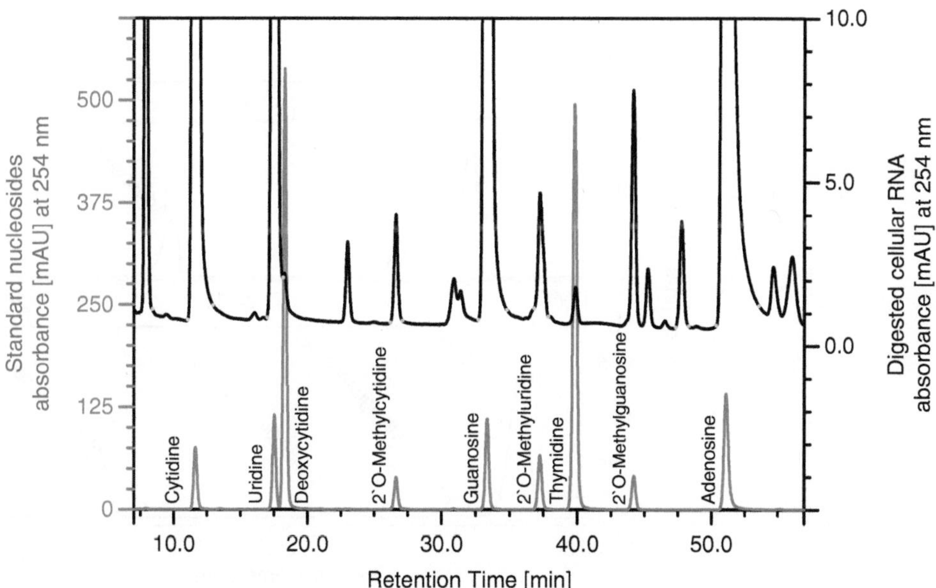

Fig. 1 Chromatogram overlay of standard nucleosides cytidine, uridine, 2′-deoxycytidine, 2′-O-methylcytidine, guanosine, 2′-O-methyluridine, thymidine, 2′-O-methylguanosine, and adenosine (*grey*) and nucleosides recovered from a total RNA preparation of the epithelial colorectal adenocarcinoma cell line Caco-2 (*black*). Separation was performed at a linear gradient of 0–20 % buffer B over 50 min with a flow rate of 0.85 ml/min. Column temperature was set to 21 °C

3.1.1 Standard Protocol

1. Set column oven temperature to 21 °C.
2. Set flow rate to 1 ml/min.
3. Equilibrate column with 1 column void volume of buffer A.
4. Decrease flow rate to 0.85 ml/min.
5. Load column with standard nucleosides or digested and dephosphorylated RNA (*see* Subheading 3.2).
6. Run a linear gradient from 0 to 25 % buffer B over 50 min.
7. Run 100 % buffer B for 3 column void volumes and decrease flow rate to 0.5 ml/min within 2 min in parallel.
8. Equilibrate with 3 column void volumes of buffer A and increase flow rate to 1 ml/min within 8 min.

3.1.2 Modified Protocol to Separate Nucleosides with Similar Characteristics

The separation of modified nucleosides can be limited when the nucleoside characteristics are very similar (e.g., 2′-O-methylguanosine and N1-methylguanosine). Therefore, the HPLC run conditions have to be optimized. By varying temperature and gradient slope it is possible to efficiently separate 2′-O-methylguanosine and N1-methylguanosine which is not achieved by the standard protocol (Fig. 2) (*see* **Note 8**).

1. Set column oven to 8 °C.
2. Set flow rate to 0.85 ml/min.

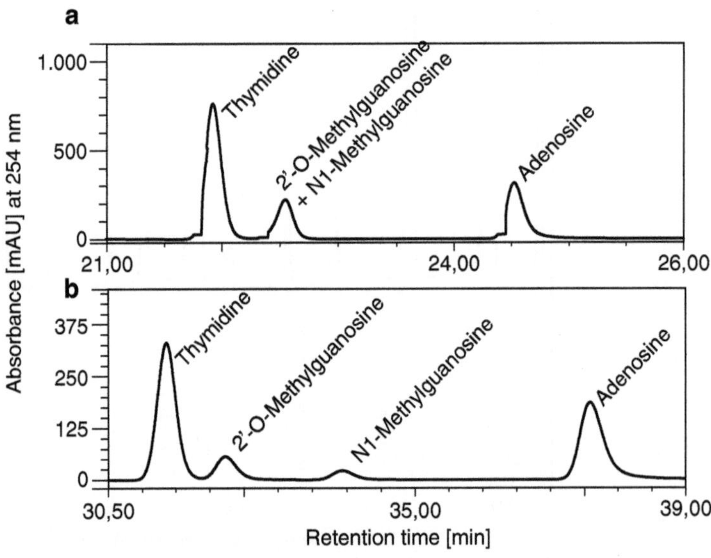

Fig. 2 (**a**) Chromatogram with the standard nucleosides thymidine, 2′-O-methylguanosine, N1-methylguanosine, and adenosine. The gradient was set from 0 to 60 % buffer B over 66.5 min. Column temperature was set to 37 °C. (**b**) Chromatogram with the standard nucleosides from (**a**) with a buffer B gradient from 7 to 17.4 % over 35 min and a flow rate of 0.85 ml/min. Column temperature was set to 8 °C

3. Equilibrate column with 1 column void volume of buffer A.
4. Load column with selected standard nucleosides or digested and dephosphorylated RNA.
5. Run a linear gradient from 7 to 17.4 % buffer B over 35 min.
6. Decrease flow rate to 0.5 ml/min.
7. Run 100 % buffer B for 3 column void volumes.
8. Equilibrate with 3 column void volumes of buffer A.

3.1.3 UV Spectra for Discrimination of Nucleosides

The nucleoside specific UV spectra are an additional characteristic that can be used to discriminate nucleosides. For example, adenosine, guanosine, and 8-hydroxyguanosine differ in the wavelength of maximum absorbance and number of peaks (Fig. 3a).

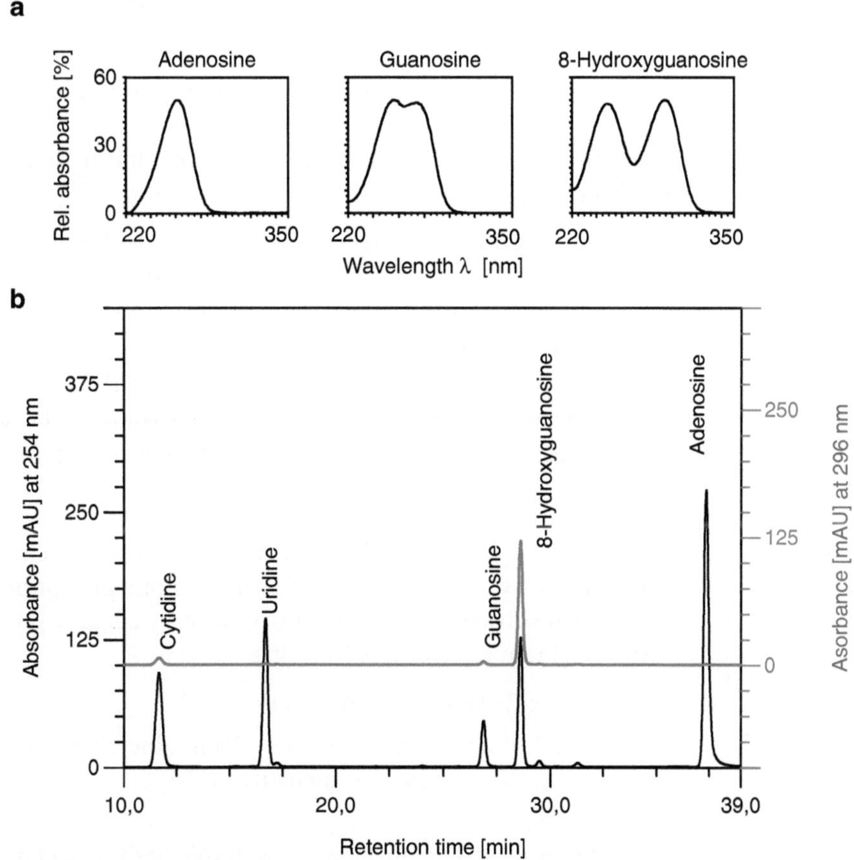

Fig. 3 (**a**) UV-spectrum of adenosine with one distinct absorption maximum at 261.0 nm, of guanosine with two nearby relative maxima at 255.9 and 274.0 nm and of 8-hydroxyguanosine with maxima at 248.2 and 294 nm. (**b**) Chromatogram overlay of nucleosides containing a mixture of guanosine, cytidine, uridine, adenosine, and 8-hydroxyguanosine. Absorbance at 254 nm (*black*) and 296 nm (*grey*) were detected in parallel. The column temperature was set to 21 °C and separation was performed at a linear gradient of 0–25 % buffer B 50 min with a flow rate of 0.85 ml/min

A suitable UV detector connected to the HPLC system allows to record UV spectra and retention time simultaneously. Using the following protocol 8-hydroxyguanosine, which has two absorbance maxima at 248 and 294 nm, can be distinguished from guanosine by retention time and UV-absorbance (Fig. 3b).

1. Set column oven to 21 °C.
2. Set flow rate to 1 ml/min.
3. Equilibrate column with 1 column void volume of buffer A.
4. Set flow rate to 0.85 ml/min.
5. Load column with selected standard nucleosides or digested and dephosphorylated RNA.
6. Run a linear gradient from 0 to 20 % buffer B over 50 min.
7. Decrease flow rate to 0.5 ml/min.
8. Run 100 % buffer B for 3 column void volumes.
9. Increase flow rate to 1 ml/min.
10. Equilibrate with 3 column void volumes of buffer A.

3.1.4 Quantification of Nucleosides

Quantification can be achieved using the linear standard plot method [28].

1. Dilute 500 μM of the relevant nucleoside 1:1 for ten times. Subject these standards to HPLC analysis and plot the values of the peak areas against the concentration to create a linear standard curve.
2. The software Chromeleon 6.80 SR10 Build 2818 can determine the peak area of the relevant nucleoside in the sample of interest and calculate the absolute amount using the standard curve.

3.2 RNA Digestion

For HPLC analysis the RNA of interest has to be cleaved into nucleotides by nuclease P1 and snake venom phosphodiesterase (SvP) with subsequent dephosphorylation by alkaline phosphatase (AP) to obtain nucleosides [29, 30].

1. Dissolve 2–100 μg RNA in 60 μl H_2O.
2. Add 10 μl ammonium acetate, 20 μl zinc chloride, and 5 μl of P1 endonuclease and incubate at 37 °C overnight or 42 °C for 2 h.
3. Add 15 μl of Tris buffer and 15 μl magnesium acetate.
4. Use 2.5 μl SvP and 0.25 μl AP and incubate at 37 °C for 2 h. After dephosphorylation, centrifuge at 30,000 × *g* for 10 min and harvest the supernatant carefully. Adjust the volume to 130 μl with buffer A and use for HPLC injection.

3.3 Competitive 8-Hydroxyguanosine ELISA

The unknown 8-OHG samples or 8-OHG standards are added to an 8-OHG/BSA conjugate preabsorbed microtiter plate. Then an anti-8-OHG monoclonal antibody is added with subsequent detection by a secondary peroxidase-labeled antibody. The 8-OHG content in unknown samples is determined by comparison with the 8-OHG standard curve [31].

1. Dilute 8-OHG-BSA conjugate 1:200,000 in PBS. Coat each well of the microtiter plate with 50 μl. Store at 4 °C overnight.
2. Wash each well six times with 250 μl washing buffer.
3. Block plates with 250 μl/well blocking buffer 2. Incubate for 2 h at RT and wash six times.
4. Prepare the RNA in a twofold dilution series six times starting with 20 μg/ml in PBS. Dilute the 8-OHG standard 1:1 in PBS for 12 times starting at 2 μg/ml. Add 50 μl of standards and samples to the wells. Use three replicates for each sample and duplicates for standard (*see* **Note 9**).
5. Immediately add 50 μl of 8-OHG antibody diluted 1:10,000 in blocking buffer 1 to each well. Incubate for 1 h at RT.
6. Wash the plate six times and add 50 μl of peroxidase-conjugated goat anti-mouse IgG diluted 1:20,000 in blocking buffer 1. Incubate for 1 h at RT.
7. Wash the plate six times and add 50 μl substrate solution per well. Incubate for approximately 30 min in the dark and terminate the reaction by adding 25 μl 2 M H_2SO_4 to each well.
8. The absorbance is measured at 490 nm against 650 nm used as reference wavelength.

4 Notes

1. For preparing buffer A dissolve 0.385 g of ammonium acetate in 50 ml H_2O and adjust pH value with 10 % glacial acetic acid to pH 6.0. To sterilize solution use a 0.2 μm disposable filter (Kobe, Germany) and filtrate into 950 ml HPLC grade H_2O.
2. For all buffers HPLC gradient grade solution should be used. Adhere to professional storage, handling, and disposal of buffers. Buffers can be stored at room temperature. All chemicals used for buffers were supplied from Carl Roth, Germany.
3. Guanosine and methylated derivatives should be dissolved in buffer A whereas adenosine and methylated derivatives are more soluble in H_2O. Solubility is enhanced by increasing temperature up to 50 °C and ultrasonic treatment. Other nucleosides can be dissolved in H_2O or buffer A. For long-term storage, aliquot nucleosides and freeze at −20 °C. To set up working

solutions, thaw aliquots and mix nucleosides in buffer A to yield working concentrations listed in Subheading 2.2.

4. All buffers for RNA digestion should be sterilized by filtration (0.2 μm filter) and stored at 4 °C. Chemicals were supplied from Carl Roth, Germany.
5. Total cellular or in vitro transcribed RNA can be used, but should be phenol–chloroform-purified before digestion and HPLC analysis. RNA should be stored at −80 °C.
6. All chemicals were supplied by Carl Roth, Germany or as indicated. BSA conjugated 8-OHG was generated as described by Senapathy et al. [32]. Store conjugated protein at −20 °C.
7. HPLC equilibration, separation, and cleaning procedure:
   (a) Set up equilibration: Rinse the column which is stored in buffer C with at least 5 void volumes of buffer C, buffer B and buffer A to equilibrate the column. Take care to remove air bubbles thoroughly before connecting the column.
   (b) Separation: Set column oven temperature and run parameters according to the protocols in Subheading 3.
   (c) Cleaning procedure: Wash column with least 5 void volumes of buffer B followed by the equal amount of buffer C. If the column is not used for more than 1 week, remove and seal it.
8. Change parameters of the standard protocol in the order given to optimize HPLC run such as gradient slope, flow rate, column temperature, eluotropic strength of elution buffer, ion strength of buffer A, pH value of buffer A, use of multidimensional buffer systems [33].
9. For some RNA samples digestion and dephosphorylation before 8-OHG detection may increase sensitivity.

Acknowledgments

These studies were funded by the German research foundation (DFG) (Grant BA 1618/5-1) and the von Behring Röntgen Stiftung (Grant 56-0034).

References

1. Limbach PA, Crain PF, McCloskey JA (1994) Summary: the modified nucleosides of RNA. Nucleic Acids Res 22:2183–2196
2. Bachellerie JP, Cavaille J (1997) Guiding ribose methylation of rRNA. Trends Biochem Sci 22:257–261
3. Bjork GR (1975) Transductional mapping of gene trmA responsible for the production of 5-methyluridine in transfer ribonucleic acid of Escherichia coli. J Bacteriol 124:92–98
4. Bjork GR, Ericson JU, Gustafsson CE, Hagervall TG, Jonsson YH, Wikstrom PM

(1987) Transfer RNA modification. Annu Rev Biochem 56:263–287

5. De Bie LG, Roovers M, Oudjama Y, Wattiez R, Tricot C, Stalon V, Droogmans L, Bujnicki JM (2003) The yggH gene of Escherichia coli encodes a tRNA (m7G46) methyltransferase. J Bacteriol 185:3238–3243
6. Persson BC, Jager G, Gustafsson C (1997) The spoU gene of Escherichia coli, the fourth gene of the spoT operon, is essential for tRNA (Gm18) 2′-O-methyltransferase activity. Nucleic Acids Res 25:4093–4097
7. Rottman FM, Bokar JA, Narayan P, Shambaugh ME, Ludwiczak R (1994) N6-adenosine methylation in mRNA: substrate specificity and enzyme complexity. Biochimie 76:1109–1114
8. Dominissini D, Moshitch-Moshkovitz S, Schwartz S, Salmon-Divon M, Ungar L, Osenberg S, Cesarkas K, Jacob-Hirsch J, Amariglio N, Kupiec M, Sorek R, Rechavi G (2012) Topology of the human and mouse m^6A RNA methylomes revealed by m^6A-seq. Nature 485:201–206
9. Meyer KD, Saletore Y, Zumbo P, Elemento O, Mason CE, Jaffrey SR (2012) Comprehensive analysis of mRNA methylation reveals enrichment in 3′ UTRs and near stop codons. Cell 149:1635–1646
10. Kohchi C, Inagawa H, Nishizawa T, Soma G (2009) ROS and innate immunity. Anticancer Res 29:817–821
11. Tschopp J, Schroder K (2010) NLRP3 inflammasome activation: the convergence of multiple signalling pathways on ROS production? Nat Rev Immunol 10:210–215
12. Pacher P, Beckman JS, Liaudet L (2007) Nitric oxide and peroxynitrite in health and disease. Physiol Rev 87:315–424
13. Barciszewski J, Barciszewska MZ, Siboska G, Rattan SI, Clark BF (1999) Some unusual nucleic acid bases are products of hydroxyl radical oxidation of DNA and RNA. Mol Biol Rep 26:231–238
14. Shan X, Chang Y, Lin CL (2007) Messenger RNA oxidation is an early event preceding cell death and causes reduced protein expression. FASEB J 21:2753–2764
15. Nunomura A, Moreira PI, Castellani RJ, Lee HG, Zhu X, Smith MA, Perry G (2012) Oxidative damage to RNA in aging and neurodegenerative disorders. Neurotox Res 22:231–248
16. Takeuchi O, Akira S (2010) Pattern recognition receptors and inflammation. Cell 140:805–820
17. Blasius AL, Beutler B (2010) Intracellular Toll-like receptors. Immunity 32:305–315
18. Kato H, Takeuchi O, Sato S, Yoneyama M, Yamamoto M, Matsui K, Uematsu S, Jung A, Kawai T, Ishii KJ, Yamaguchi O, Otsu K, Tsujimura T, Koh CS, Reis e Sousa C, Matsuura Y, Fujita T, Akira S (2006) Differential roles of MDA5 and RIG-I helicases in the recognition of RNA viruses. Nature 441:101–105
19. Hornung V, Ellegast J, Kim S, Brzozka K, Jung A, Kato H, Poeck H, Akira S, Conzelmann KK, Schlee M, Endres S, Hartmann G (2006) 5′-Triphosphate RNA is the ligand for RIG-I. Science 314:994–997
20. Pichlmair A, Schulz O, Tan CP, Naslund TI, Liljestrom P, Weber F, Reis e Sousa C (2006) RIG-I-mediated antiviral responses to single-stranded RNA bearing 5′-phosphates. Science 314:997–1001
21. Pichlmair A, Schulz O, Tan CP, Rehwinkel J, Kato H, Takeuchi O, Akira S, Way M, Schiavo G, Reis e Sousa C (2009) Activation of MDA5 requires higher-order RNA structures generated during virus infection. J Virol 83:10761–10769
22. Kariko K, Buckstein M, Ni H, Weissman D (2005) Suppression of RNA recognition by Toll-like receptors: the impact of nucleoside modification and the evolutionary origin of RNA. Immunity 23:165–175
23. Hamm S, Latz E, Hangel D, Muller T, Yu P, Golenbock D, Sparwasser T, Wagner H, Bauer S (2010) Alternating 2′-O-ribose methylation is a universal approach for generating non-stimulatory siRNA by acting as TLR7 antagonist. Immunobiology 215:559–569
24. Jockel S, Nees G, Sommer R, Zhao Y, Cherkasov D, Hori H, Ehm G, Schnare M, Nain M, Kaufmann A, Bauer S (2012) The 2′-O-methylation status of a single guanosine controls transfer RNA-mediated Toll-like receptor 7 activation or inhibition. J Exp Med 209:235–241
25. Robbins M, Judge A, Liang L, McClintock K, Yaworski E, MacLachlan I (2007) 2′-O-methyl-modified RNAs act as TLR7 antagonists. Mol Ther 15:1663–1669
26. Daffis S, Szretter KJ, Schriewer J, Li J, Youn S, Errett J, Lin TY, Schneller S, Zust R, Dong H, Thiel V, Sen GC, Fensterl V, Klimstra WB, Pierson TC, Buller RM, Gale M Jr, Shi PY, Diamond MS (2010) 2′-O methylation of the viral mRNA cap evades host restriction by IFIT family members. Nature 468:452–456
27. Zust R, Cervantes-Barragan L, Habjan M, Maier R, Neuman BW, Ziebuhr J, Szretter KJ, Baker SC, Barchet W, Diamond MS, Siddell SG, Ludewig B, Thiel V (2011) Ribose 2′-O-methylation provides a molecular signature for the distinction of self and non-self

mRNA dependent on the RNA sensor Mda5. Nat Immunol 12:137–143

28. Li D, Martini N, Wu Z, Wen J (2012) Development of an isocratic HPLC method for catechin quantification and its application to formulation studies. Fitoterapia 83:1267–1274
29. Fujimoto M, Kuninaka A, Yoshino H (1974) Identity of phosphodiesterase and phosphomonoesterase activities with nuclease P1. Agric Biol Chem 38:785–790
30. Williams EJ, Sung SC, Laskowski M Sr (1961) Action of venom phosphodiesterase on deoxyribonucleic acid. J Biol Chem 236:1130–1134
31. Chiou CC, Chang PY, Chan EC, Wu TL, Tsao KC, Wu JT (2003) Urinary 8-hydroxydeoxyguanosine and its analogs as DNA marker of oxidative stress: development of an ELISA and measurement in both bladder and prostate cancers. Clin Chim Acta 334:87–94
32. Senapathy P, Ali MA, Jacob MT (1985) Mechanism of coupling periodate-oxidized nucleosides to proteins. FEBS Lett 190:337–341
33. Snyder LR (1997) Changing reversed-phase high performance liquid chromatography selectivity. Which variables should be tried first? J Chromatogr B Biomed Sci Appl 689:105–115

Chapter 2

Enzymatic Synthesis and Purification of a Defined RIG-I Ligand

Marion Goldeck, Martin Schlee, Gunther Hartmann, and Veit Hornung

Abstract

Receptor-based nucleic acid sensing constitutes one of the most fundamental mechanisms of our innate immune system to sense viral infection. RIG-I is a cytosolic RNA helicase that senses the presence of 5′ triphosphate RNA species, a common feature of many negative strand RNA viruses. We here describe a protocol to enzymatically synthesize and to purify a defined RIG-I ligand that can be used to study RIG-I activation in vitro and in vivo.

Key words 5-Triphosphate RNA, RIG-ligand, Interferon induction, RNA viruses

1 Introduction

The innate immune system comprises a network of pathogen detecting receptors, which trigger intracellular defense mechanisms and the induction of innate immune responses against invading pathogens [1]. Measures against infecting viruses include inhibition of translation, RNA degradation, induction of apoptosis, and secretion of cytokines and chemokines, leading to alarming of neighboring cells and attracting immune cells [2]. In addition, the stimulation of innate immune receptors is crucial for initiation of an adaptive immune response [3]. The cytosolic immune receptor RIG-I is important for the immune response against RNA viruses [4]. It detects viruses by recognition of viral 5′ triphosphorylated RNA [5, 6]. Unlike other nucleic acid sensing innate immune receptors (e.g., TLR7, 8, or 9) RIG-I is ubiquitously expressed [7], even in transformed tumor cells [8, 9]. The fact that most of the highly pathogenic and emerging viruses are sensed by RIG-I (e.g., Influenza) highlights the relevance of this pathway. Moreover, in addition to pro-inflammatory cytokine induction, RIG-I can also trigger apoptosis [10]. By transfection of RIG-I stimulating ligands, which mimic a viral infection in the absence of

Hans-Joachim Anders and Adriana Migliorini (eds.), *Innate DNA and RNA Recognition*, Methods in Molecular Biology, vol. 1169, DOI 10.1007/978-1-4939-0882-0_2, © Springer Science+Business Media New York 2014

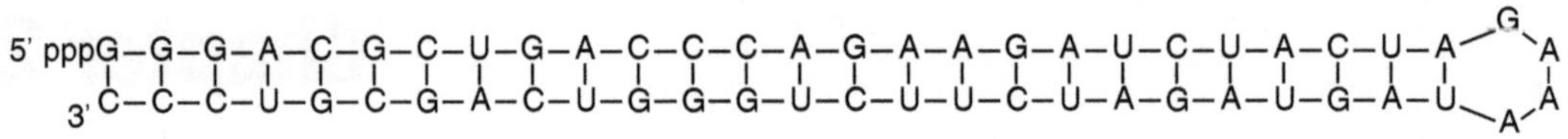

```
                                                                                     G
5' pppG–G–G–A–C–G–C–U–G–A–C–C–C–A–G–A–A–G–A–U–C–U–A–C–U–A    A
      | | | | | | | | | | | | | | | | | | | | | | | | | |    |
3'    C–C–C–U–G–C–G–A–C–U–G–G–G–U–C–U–U–C–U–A–G–A–U–G–A–U    A
                                                                                     A
```

Fig. 1 The predicted secondary structure of IVT4

a virus, the RIG-I induced innate immune response can be exploited against tumor cells [8–11] and chronic or establishing viral infections [12–14] or for regulation of uncontrolled immune responses [15]. Moreover, employing defined RIG-I ligands can be advantageous for the analysis of pure RIG-I immune responses in absence of viral immune modulatory molecules [16].

RIG-I was found to recognize 5′ triphosphorylated RNA (pppRNA) [5, 6]. In a follow-up study, using synthetic pppRNA, we further dissected the RIG-I activating RNA ligand structure [17] (*see* Fig. 1): We found that RIG-I activation requires a base-paired 5′ end, which has to be part of an at least 19–20 nt base-paired RNA stretch. While blunt-ended, base-paired pppRNA (ppp-dsRNA) represents an optimal ligand for RIG-I activation, 3′ overhangs reduce and 5′ overhangs at the ppp-end abolish RIG-I activation. Indeed, crystal structure data explain the necessity of a based paired 5′ end: The base pairing at the 5′ ppp end supports a crucial stacking interaction with a phenylalanine residue in the RNA binding cleft of the C-terminal domain of RIG-I [18, 19]. In addition, in contrast to unpaired nucleotides, which allow free rotation of the following sequence, the base-paired RNA assembly stabilizes the helix in an optimal position for interaction of the adjoining phosphodiester backbone with the C-terminal domain and the helicase domain. This interaction is the prerequisite for RIG-I activation (reviewed in ref. 20). Since phage polymerase in vitro transcription (IVT) generates side products complementary to the intended transcribed sequence, any in vitro transcribed RNA (mixture), even if designed as single strand, will exhibit some RIG-I stimulating activity [17, 21]. For this reason, in previous or even in some recent studies "single stranded" triphosphorylated RNA is still considered as RIG-I ligand. However, the amount of active ligand in such in vitro transcribed "single stranded" RNA mixtures is expected to be very low and uncontrolled. To this end, we here provide a protocol to generate highly active ppp-dsRNA in a two-step protocol employing an enzymatic in vitro transcription reaction, followed by a PAGE-based purification procedure. For simplification and standardization purposes, this protocol relies on the enzymatic synthesis of a single, self-annealing hairpin RNA oligonucleotide that forms a strong intramolecular hairpin structure, thus automatically forming a dsRNA molecule (IVT4). With appropriate complexation, this molecule can be delivered into cells as a defined RIG-I stimulus for in vitro and in in vivo studies [9, 11, 16, 17, 22, 23].

2 Materials

Prepare all solutions using sterile, nuclease-free water and analytical grade reagents. Make sure to avoid any contamination by RNases by using clean, RNase-decontaminated flasks, spatulas, and pipettes. Prepare and store all solutions at room temperature if not specified otherwise. Store the in vitro transcription enzymes and RNA monomers and oligonucleotides at −20 °C.

2.1 In Vitro Transcription

2.1.1 Annealing of the IVT DNA Template

1. 10× Hybridization buffer (250 mM Tris–HCl, 250 mM NaCl, pH 7.4).
2. RNase-free H_2O.
3. IVT4 Sense Oligo (5′-TTGTAATACGACTCACTATAGGG ACGCTGACCCAGAAGATCTACTAGAAATAGT AGATCTTCTGGGTCAGCGTCCC-3′) 100 μM.
4. IVT4 Antisense Oligo (5′-GGGACGCTGACCCAGAAGATC TACTATTTCTAGTAGATCTTCTGGGTCAGCGTC CCTATAGTGAGTCGTATTACAA-3′) 100 μM.

2.1.2 In Vitro Transcription of IVT4 5′-ppp-dsRNA

1. 5× Reaction buffer (provided together with the RNA polymerase chosen).
2. T7 RNA Polymerase Enzyme Mix.
3. RNase-free H_2O.
4. 100 μM ATP, 100 μM CTP, 100 μM GTP, 100 μM UTP solution.
5. IVT DNA template (*see* above Subheading 2.1.1).
6. DNase I, RNase free (1 U/μl).
7. Mini Quick Spin Columns for DNA/RNA or oligonucleotides purification.
8. 0.2 and 2.0 ml plastic centrifuge tubes.

2.2 Polyacrylamide Gel Components

1. RNase decontamination solution.
2. 30 % Acrylamide–Bisacrylamide solution (37.5:1, Carl Roth, *see* **Note 1**).
3. Tris–borate–EDTA (10× TBE): 890 mM Tris, pH 8.3, 890 mM boric acid, 20 mM EDTA.
4. Urea.
5. 10 % Ammonium peroxodisulfate solution (APS): Dissolve 1 g APS in 10 ml water. Store at 4 °C (*see* **Note 2**).
6. *N*,*N*,*N*′,*N*′-Tetramethylethylenediamine (TEMED). Store at 4 °C (*see* **Note 3**).
7. TBE electrophoresis buffer (1×): 89 mM Tris–HCl, 89 mM boric acid, 2 mM EDTA. Add 100 ml of 10× TBE buffer to 900 ml of water and mix well.

8. Loading buffer (2.5×): 222.5 mM Tris–HCl, 222.5 mM boric acid, 5 mM EDTA, ~0.1 % Orange G, 75 % formamide. Mix 7.5 ml of formamide RNase/DNase free and 2.5 ml of 10× TBE and add a spatula tip of Orange G (Sigma-Aldrich).
9. DNA ladder: Gene Ruler ULR DNA Ladder, 0.5 μg/μl.
10. RNA standards: 27mer RNA (5′-GACGCUGACCCUGAAGUUCAUCUUACG-3′), 48mer RNA (5′-GGGACUGACCCUGAAGUUCAUCUUCCGACAUACUUCCGUGUACGUCCC-3′).

2.3 Gel Extraction Components

1. 0.5, 1, and 2 ml sterile plastic centrifuge tubes.
2. Scalpel.
3. Silica-covered TLC plate (Merck, TLC silica gel 60 F_{254}).
4. Short-wavelength (254 nm) UV light source.
5. 5 M NaCl solution in water.
6. 100 % Ethanol.

3 Methods

3.1 In Vitro Transcription

3.1.1 Annealing of the IVT DNA Template

1. Mix 3 μl 10× hybridization buffer, 21 μl RNase-free H_2O, 3 μl IVT4 Sense Oligo, 3 μl IVT4 Antisense Oligo in a 0.2 ml PCR reaction tube.
2. Heat above mixture for 5 min at 90 °C, then cool to 45 °C at −1 °C/min, e.g., using a standard PCR thermocycler.
3. The IVT DNA template can be stored at −20 °C or directly used in the following IVT reaction.

3.1.2 In Vitro Transcription of IVT4 5′-ppp-dsRNA

1. Combine 6 μl H_2O, 12 μl 5× reaction buffer, 12 μl IVT DNA template, 6 μl ATP, 6 μl GTP, 6 μl UTP, 6 μl CTP, and 6 μl T7 RNA Polymerase Enzyme Mix in a 0.5 ml reaction tube (final volume 60 μl).
2. Incubate at 37 °C for 6 h.
3. Add 6 μl DNase I.
4. Incubate at 37 °C for 30 min.
5. Apply above mixture (66 μl) to a pre-centrifuged Mini Quick Spin DNA/RNA Column that is positioned in a 2.0 ml reaction tube to collect the flow through.
6. Centrifuge at 1,000 ×*g* for 4 min at RT.
7. The flow through fraction contains the RNA in 10 mM Tris–HCl, pH 7.5, 1 mM EDTA, 100 mM NaCl in approximately 66 μl (*see* **Note 4**).

3.2 Preparative 15 % Polyacrylamide Gel Electrophoresis

The following PAGE purification procedure is optional. In our experience, the column-purified RNA is ready for subsequent in vitro or in vivo applications, despite slight contaminations. However, to guarantee a high inter-batch consistency, we recommend subjecting the IVT reaction to the following PAGE purification procedure.

1. Mix 7.5 ml of 30 % Acrylamide–Bisacrylamide solution and 1.5 ml of TBE (10×) in a 50 ml centrifuge tube and add 7.2 g urea. Incubate in a water bath at about 37 °C and stir occasionally until obtaining a homogenous solution. Add 150 μl of APS and 12 μl of TEMED and mix briefly. Pour the gel mixture into a 9 cm×9 cm×1.5 mm gel cassette and insert a 15-well comb immediately. Let it polymerize (*see* **Note 5**).
2. Attach the gel cassette to the electrophoresis cell, fill the buffer chamber with TBE (1×), and pre-run the gel for 20 min at 120 V.
3. Prepare samples of 30–60 μg crude in vitro transcripts per purification by diluting 3 sample volumes with 1 volume of loading buffer. For purification of higher amounts up to three purifications may be loaded on one gel. Heat the sample to 90 °C for 3 min and then quickly snap-cool on ice (*see* **Note 6**). Briefly centrifuge the samples to bring down the condensate. Do not heat the DNA ladder (*see* **Note 7**). Rinse the wells with buffer to remove any urea directly before loading. Distribute the in vitro transcript sample among three wells and load the DNA ladder (1 μg/lane) in a one-lane distance. Electrophorese at 150 V for approximately 1.5 h until the dye front has reached the bottom of the gel.

3.3 RNA Visualization by UV-Shadowing and RNA Extraction

1. Separate the glass plates of the gel cassette and carefully place the gel between plastic foils (*see* **Note 8**). Place the wrapped gel on top of the silica-coated side of the TLC plate and hold the UV light source over the gel. Nucleic acids will cast a shadow on the TLC plate and be visualized as bright purple bands (*see* **Note 9**).
2. Use a sterile scalpel to excise the product containing gel fragments. In adaptation to the number of purifications loaded cut out one or multiple fragments of about 15 mm×0.3 mm in size including the product bands three lanes (*see* **Note 10**).
3. Place a nuclease-free 0.5 ml tube into a 1.5 ml tube, perforate the bottom of the 0.5 ml tube with a sterile needle, and transfer the gel fragment into the 0.5 ml tube. Centrifuge at 10,000×*g* for 2 min to crush the gel fragment and collect a fine-grained gel pulp within the 1.5 ml tube. Repeat the centrifugation step in case any gel portions remain in the 0.5 ml tube (*see* **Note 11**).

4. Discard the 0.5 ml tube, add 400 μl of TBE buffer (1×) and incubate with vigorous shaking at 37 °C for 4 h. Cool the samples briefly on ice and centrifuge at 5,000 × *g* for 30 s.
5. Prepare a filter unit by cutting the end of a sterile 1 ml pipette tip and placing it into a 2 ml tube. Transfer the gel suspension to the filter unit with a 200 μl sterile pipette tip with the end cut. Centrifuge at 3,000 × *g* for 2 min to collect the product-containing solution within the 2 ml tube.
6. Add 24 μl of sodium chloride and 1.2 ml of ethanol to precipitate the product RNA. Incubate at −20 °C overnight. Centrifuge at 12,000 × *g* for 12 min, wash the pellet with 200 μl of ice-cold ethanol, and centrifuge again at 12,000 × *g* for 6 min. Discard the supernatant, let the pellet dry for about 15 min, and dissolve the pellet in 20 μl of water.
7. Measure the absorbance at 260 nm to quantify the purified probe (*see* **Note 12**).
8. The pppRNA is now ready for in vitro or in vivo delivery.

4 Notes

1. Unpolymerized acrylamide is a potent neurotoxin. The acrylamide–bisacrylamide solution should be handled with great care, and skin contact should be avoided.
2. In our hands the APS solution can be used for a period of 3–4 weeks without any problems.
3. TEMED is a caustic reagent that should be handled with gloves. The reagent should be pipetted quickly and stored at 4 °C to avoid exposure to its strong amine smell.
4. Gel filtration purification of the crude in vitro transcript reaction mixture using Mini Quick Spin RNA/DNA columns allows partial removal of excess reaction buffer components, nucleotides and short run-off products ≤20 base pairs. Routinely, RNA concentrations of 1–2 μg/μl are obtained after this first purification step.
5. Before assembling the gel cassette treat the glass plates, spacers and combs with RNase decontamination solution, rinse with ethanol and dry with a clean paper towel. The bottom of the gel cassette was sealed using 0.5 % agarose. The polymerization reaction was tracked by excess gel mixture in the 50 ml centrifuge tube and quantitative polymerization was ensured by an additional reaction time of 15 min.
6. Heating of the samples in the presence of formamide will denature secondary structures present within the RNA in vitro transcripts. Snap-cooling thereby prevents renaturation before

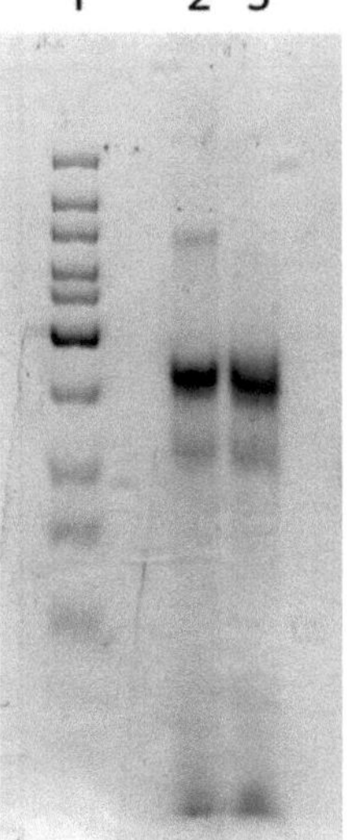

Fig. 2 Analytical native PAGE gel (15 %) showing the effects of snap-cooling during sample preparation. *Lane 1* shows Gene Ruler ULR DNA Ladder. *Lanes 2* and *3* show IVT4 untreated and pretreated by snap-cooling, respectively. 1 μg RNA was loaded on each lane. The gel was stained with methylene blue

the samples are loaded onto the gel. In the case of the IVT4 sequence being designed to form a self-complementary hairpin this pretreatment shifts the hairpin-duplex equilibrium from the bimolecular duplex toward the intermolecular assembly. Especially undefined dimers or multimers containing transcription byproducts are dissolved helping to obtain a single well-defined IVT4 product band. This beneficial effect of the sample pretreatment can be followed by native PAGE analysis (Fig. 2).

7. Avoiding heating under denaturing conditions helps to maintain the integrity of the DNA ladder duplexes. Note that comparison of the resulting in vitro transcript bands with the marker bands does not allow size determination of the products as the running behavior differs from DNA to RNA and is sequence-specific. Loading of the DNA ladder especially serves to ensure accurate electrophoretic separation and to compare results of different runs. Additional loading of purified IVT4 as a known size standard or synthetic dsRNA of similar length and size will help to identify the appropriate product band within the in vitro reaction mixture.
8. Treat the plastic foils with RNase decontamination solution prior use and avoid contact with the RNA containing gel areas while transferring the gel to the plastic foil. We have made good experience by using plastic disposal bags (BRAND) for wrapping the gel.

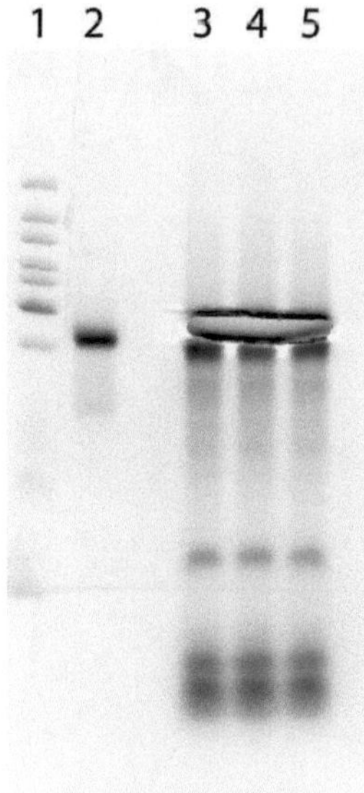

Fig. 3 Preparative denaturing PAGE gel (15 %) after excision of the IVT4 product band. *Lane 1* shows Gene Ruler ULR DNA Ladder. *Lane 2* shows PAGE-purified IVT4 product (1 μg) loaded as a known size standard. *Lanes 3–5* show the remaining byproducts of a IVT4 in vitro transcription (30 μg) after excision of the product band. For visualization of the bands the PAGE gel was stained with methylene blue after isolation of the desired gel fragment

9. Be aware that long wavelength UV light will not work. UV-shadowing allows visualization of RNA with a detection limit of about 0.2 μg/band.
10. During excision of the RNA product band exposure of the gel to UV light should be kept at a minimum to avoid UV-induced RNA damage. We find it helpful to first mark the product band on the covering plastic foil under UV light, then excise the band with the scalpel and subsequently ensure successful excision by re-submission to UV light. The product bands should be cut out as small as possible (Fig. 3). We recommend that gel fragments exceeding a size of 15 mm × 0.3 mm should be partitioned for elution to ensure efficient recovery of the RNA.
11. The surface of the gel fragment is significantly increased by this procedure accelerating the subsequent elution by diffusion. Alternatively, directly transfer the gel fragment into a 1.5 ml centrifuge tube and use a sterile pipette tip to crush the gel against the wall of the tube.
12. PAGE purification of 30 μg crude in vitro transcript commonly yields about 7–10 μg IVT4 RNA leading to 30–50 μg IVT4 RNA for purification of an entire IVT reaction. Significant amounts of short run-off by-products that are still present after

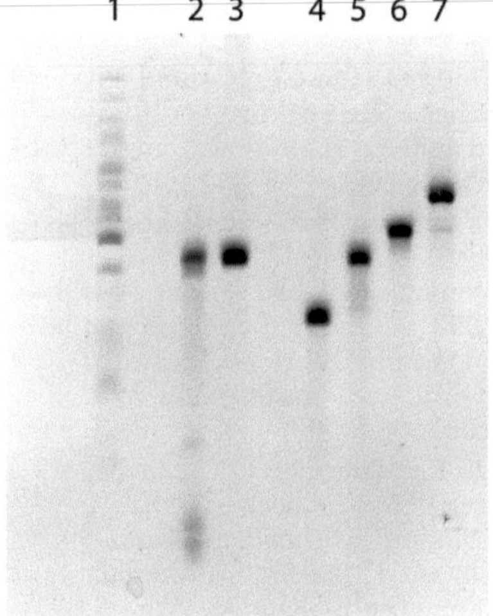

Fig. 4 Analytical denaturing PAGE gel (15 %) of IVT4 gel purification. *Lane 1* shows Gene Ruler ULR DNA Ladder. *Lane 2* and *3* show IVT4 before and after gel purification, respectively. *Lanes 4–7* show RNA standards: 27mer ssRNA (*4*), 27mer dsRNA (*5*), 48mer ssRNA (*6*), 48mer dsRNA (*7*). 1 μg RNA was loaded on each lane. The gel was stained with methylene blue

the first gel filtration purification step are successfully removed yielding a single, well-defined IVT4 product band in denaturing PAGE analysis (Fig. 4). Comparison to a 27mer dsRNA standard proves the right size of the purified IVT4 hairpin.

Acknowledgments

G.H. and V.H. are supported by the excellence cluster ImmunoSensation. M.S., G.H., and V.H. are supported by grants from the German Research Foundation (SFB704 to V.H. and G.H., SFB670 to M.S., G.H., and V.H., DFG Research Grants Program SCHL 1930/1-1 to M.S.) and the European Research Council (ERC 243046 to V.H.).

References

1. Takeuchi O, Akira S (2010) Pattern recognition receptors and inflammation. Cell 140(6): 805–820. doi:10.1016/j.cell.2010.01.022
2. Sadler AJ, Williams BR (2008) Interferon-inducible antiviral effectors. Nat Rev Immunol 8(7):559–568. doi:10.1038/nri2314
3. Aoshi T, Koyama S, Kobiyama K, Akira S, Ishii KJ (2011) Innate and adaptive immune responses to viral infection and vaccination. Curr Opin Virol 1(4):226–232. doi:10.1016/j.coviro.2011.07.002
4. Schlee M, Hartmann G (2010) The chase for the RIG-I ligand: recent advances. Mol Ther 18(7):1254–1262. doi:10.1038/mt.2010.90
5. Hornung V, Ellegast J, Kim S, Brzozka K, Jung A, Kato H, Poeck H, Akira S, Conzelmann KK, Schlee M, Endres S, Hartmann G (2006) 5′-Triphosphate RNA is the ligand for RIG-I.

Science 314(5801):994–997. doi:10.1126/science.1132505

6. Pichlmair A, Schulz O, Tan CP, Naslund TI, Liljestrom P, Weber F, Reis e Sousa C (2006) RIG-I-mediated antiviral responses to single-stranded RNA bearing 5′-phosphates. Science 314(5801):997–1001. doi:10.1126/science.1132998
7. Barchet W, Wimmenauer V, Schlee M, Hartmann G (2008) Accessing the therapeutic potential of immunostimulatory nucleic acids. Curr Opin Immunol 20(4):389–395. doi:10.1016/j.coi.2008.07.007
8. Poeck H, Besch R, Maihoefer C, Renn M, Tormo D, Morskaya SS, Kirschnek S, Gaffal E, Landsberg J, Hellmuth J, Schmidt A, Anz D, Bscheider M, Schwerd T, Berking C, Bourquin C, Kalinke U, Kremmer E, Kato H, Akira S, Meyers R, Hacker G, Neuenhahn M, Busch D, Ruland J, Rothenfusser S, Prinz M, Hornung V, Endres S, Tuting T, Hartmann G (2008) 5′-Triphosphate-siRNA: turning gene silencing and Rig-I activation against melanoma. Nat Med 14(11):1256–1263. doi:10.1038/nm.1887
9. Kubler K, Gehrke N, Riemann S, Bohnert V, Zillinger T, Hartmann E, Polcher M, Rudlowski C, Kuhn W, Hartmann G, Barchet W (2010) Targeted activation of RNA helicase retinoic acid-inducible gene-I induces proimmunogenic apoptosis of human ovarian cancer cells. Cancer Res 70(13):5293–5304. doi:10.1158/0008-5472.CAN-10-0825
10. Besch R, Poeck H, Hohenauer T, Senft D, Hacker G, Berking C, Hornung V, Endres S, Ruzicka T, Rothenfusser S, Hartmann G (2009) Proapoptotic signaling induced by RIG-I and MDA-5 results in type I interferon-independent apoptosis in human melanoma cells. J Clin Invest 119(8):2399–2411. doi:10.1172/JCI37155
11. Glas M, Coch C, Trageser D, Dassler J, Simon M, Koch P, Mertens J, Quandel T, Gorris R, Reinartz R, Wieland A, Von Lehe M, Pusch A, Roy K, Schlee M, Neumann H, Fimmers R, Herrlinger U, Brustle O, Hartmann G, Besch R, Scheffler B (2013) Targeting the cytosolic innate immune receptors RIG-I and MDA5 effectively counteracts cancer cell heterogeneity in glioblastoma. Stem Cells 31(6):1064–1074. doi:10.1002/stem.1350
12. Ebert G, Poeck H, Lucifora J, Baschuk N, Esser K, Esposito I, Hartmann G, Protzer U (2011) 5′ Triphosphorylated small interfering RNAs control replication of hepatitis B virus and induce an interferon response in human liver cells and mice. Gastroenterology 141(2): 696–706. doi:10.1053/j.gastro.2011.05.001, 706 e691-693
13. Wang Y, Wang X, Li J, Zhou Y, Ho W (2013) RIG-I activation inhibits HIV replication in macrophages. J Leukoc Biol 94:337–341. doi:10.1189/jlb.0313158
14. Goulet ML, Olagnier D, Xu Z, Paz S, Belgnaoui SM, Lafferty EI, Janelle V, Arguello M, Paquet M, Ghneim K, Richards S, Smith A, Wilkinson P, Cameron M, Kalinke U, Qureshi S, Lamarre A, Haddad EK, Sekaly RP, Peri S, Balachandran S, Lin R, Hiscott J (2013) Systems analysis of a RIG-I agonist inducing broad spectrum inhibition of virus infectivity. PLoS Pathog 9(4):e1003298. doi:10.1371/journal.ppat.1003298
15. Dann A, Poeck H, Croxford AL, Gaupp S, Kierdorf K, Knust M, Pfeifer D, Maihoefer C, Endres S, Kalinke U, Meuth SG, Wiendl H, Knobeloch KP, Akira S, Waisman A, Hartmann G, Prinz M (2012) Cytosolic RIG-I-like helicases act as negative regulators of sterile inflammation in the CNS. Nat Neurosci 15(1):98–106. doi:10.1038/nn.2964
16. Asdonk T, Motz I, Werner N, Coch C, Barchet W, Hartmann G, Nickenig G, Zimmer S (2012) Endothelial RIG-I activation impairs endothelial function. Biochem Biophys Res Commun 420(1):66–71. doi:10.1016/j.bbrc.2012.02.116
17. Schlee M, Roth A, Hornung V, Hagmann CA, Wimmenauer V, Barchet W, Coch C, Janke M, Mihailovic A, Wardle G, Juranek S, Kato H, Kawai T, Poeck H, Fitzgerald KA, Takeuchi O, Akira S, Tuschl T, Latz E, Ludwig J, Hartmann G (2009) Recognition of 5′ triphosphate by RIG-I helicase requires short blunt double-stranded RNA as contained in panhandle of negative-strand virus. Immunity 31(1):25–34. doi:10.1016/j.immuni.2009.05.008
18. Lu C, Xu H, Ranjith-Kumar CT, Brooks MT, Hou TY, Hu F, Herr AB, Strong RK, Kao CC, Li P (2010) The structural basis of 5′ triphosphate double-stranded RNA recognition by RIG-I C-terminal domain. Structure 18(8):1032–1043. doi:10.1016/j.str.2010.05.007
19. Wang Y, Ludwig J, Schuberth C, Goldeck M, Schlee M, Li H, Juranek S, Sheng G, Micura R, Tuschl T, Hartmann G, Patel DJ (2010) Structural and functional insights into 5′-ppp RNA pattern recognition by the innate immune receptor RIG-I. Nat Struct Mol Biol 17(7):781–787. doi:10.1038/nsmb.1863
20. Kolakofsky D, Kowalinski E, Cusack S (2012) A structure-based model of RIG-I activation. RNA 18(12):2118–2127. doi:10.1261/rna.035949.112
21. Schmidt A, Schwerd T, Hamm W, Hellmuth JC, Cui S, Wenzel M, Hoffmann FS, Michallet MC, Besch R, Hopfner KP, Endres S,

Rothenfusser S (2009) 5′-Triphosphate RNA requires base-paired structures to activate antiviral signaling via RIG-I. Proc Natl Acad Sci U S A 106(29):12067–12072. doi:10.1073/pnas.0900971106

22. Heutinck KM, Kassies J, Florquin S, ten Berge IJ, Hamann J, Rowshani AT (2012) SerpinB9 expression in human renal tubular epithelial cells is induced by triggering of the viral dsRNA sensors TLR3, MDA5 and RIG-I. Nephrol Dial Transplant 27(7):2746–2754. doi:10.1093/ndt/gfr690

23. Karpus ON, Heutinck KM, Wijnker PJ, Tak PP, Hamann J (2012) Triggering of the dsRNA sensors TLR3, MDA5, and RIG-I induces CD55 expression in synovial fibroblasts. PLoS One 7(5):e35606. doi:10.1371/journal.pone.0035606

Chapter 3

Crystallization of Mouse RIG-I ATPase Domain: In Situ Proteolysis

Filiz Civril and Karl-Peter Hopfner

Abstract

RIG-I is a key pattern recognition receptor that recognizes cytoplasmic viral RNA. Upon ligand binding, it undergoes a conformational change that induces an active signaling conformation. However, the details of this conformational change remain elusive until high-resolution crystal structures of different functional conformations are available. X-ray crystallography is a powerful tool to study structure–function relationships, but crystallization is often the limiting step of the method. Here, we describe the in situ in-drop proteolysis of RIG-I that yielded crystals of the ATPase domain of mouse RIG-I suitable for structure determination.

Key words RIG-I, ATPase, In situ proteolysis, Crystallization, Innate immunity, Viral RNA

1 Introduction

The innate immune response is the first and essential defense against pathogenic infections. The key players of innate immunity are germ-line-encoded pattern recognition receptors (PRRs) that detect pathogen associated molecular patterns (PAMPs) [1]. The formation of PRR-PAMP complexes initiates intracellular signaling cascades that activate antiviral response by upregulating transcription of cytokines, type I Interferons, chemokines, and antimicrobial proteins [1].

RIG-I like receptors (RLRs), namely, RIG-I (retinoic acid-inducible gene I) [2], MDA5 (melanoma differentiation-associated protein 5) [3, 4], and LGP2 (laboratory of genetics and physiology 2) [5] are cytoplasmic PRRs that detect viral RNA [6]. RLRs are superfamily 2 (SF2) ATPases consisting of an RNA-dependent ATPase domain followed by a C-terminal regulatory domain (RD) [7]. RIG-I and MDA5 have additional N-terminal tandem caspase activation and recruitment domains (CARDs) which are the signaling component of RLRs [8].

Hans-Joachim Anders and Adriana Migliorini (eds.), *Innate DNA and RNA Recognition*, Methods in Molecular Biology, vol. 1169,
DOI 10.1007/978-1-4939-0882-0_3, © Springer Science+Business Media New York 2014

The optimal PAMP recognized by RLRs is base paired RNA with a 5′ blunt end [9–12] and in case of RIG-I the presence of 5′ triphosphate (5′ PPP-dsRNA) is crucial [13–15]. In an infected cell, RIG-I RD binds to 5′ PPP-dsRNA [12, 16, 17]. This interaction causes conformational changes in the protein that lead to release of the CARDs which are believed to be buried in the inactive form of the protein [6]. The exposed CARDs are now available to interact with the adaptor protein, interferon IFN-β promoter stimulator 1 (IPS1) [18], also known as virus-induced signaling adapter (VISA) [19], mitochondrial antiviral signaling (MAVS) [20], or CARD adapter inducing IFN-β (CARDIF) [21] in a ubiquitin-dependent manner [22, 23]. IPS1 is an outer mitochondrial protein that links MDA5 and RIG-I to downstream signaling cascades [18–21]. Although this model provided substantial information about the mechanism of RIG-I activation, the role of RIG-I SF2 ATPase domain remained unclear. As mutations in ATP binding and hydrolysis sites abolished RIG-I activity the ATPase domain was believed to have a vital role in viral RNA sensing [24].

To gain further insights on RIG-I activation, we decided to crystallize mouse RIG-I, which could be purified recombinantly from *E. coli* in good quantities for structural studies (*see* **Note 1**) (several mg purified protein per liter culture) and homogeneous quality [25]. Unfortunately, the full-length and the truncated versions of RIG-I lacking RD or CARDs or both failed to form suitable crystals for structure determination in our hands, possibly due to presence of highly flexible regions as small angle X-ray crystallographic data of RIG-I suggested [26].

To overcome the problem of flexible regions possibly preventing crystal packing we have used in situ proteolysis of different mouse RIG-I constructs (*see* **Note 2**). The polypeptides purified to homogeneity were mixed with trace amounts of protease subtilisin prior to crystallization setup. Subtilisin is a nonspecific protein serine endopeptidase, and thus, it attacks the peptide bond at any vulnerable location independent of the primary structure of the polypeptide. Using this method, we managed to obtain well diffracting crystals and eventually the crystal structure of RIG-I SF2 ATPase domain from a construct that consists of CARDs and SF2 ATPase (*see* **Note 3**) (a.a. 1-797) [25]. Using different constructs or different species (human and duck), other groups obtained crystals of RIG-I or in complex with nucleic acids, other functional states, and even full-length RIG-I [26–29].

Here, we explain in detail affinity, ion exchange, and size exclusion purifications of mouse RIG-I 1-797 construct (Fig. 1) yielding good quality and quantity of the polypeptide (Fig. 2a) and then its use in in situ proteolysis method with protease subtilisin to get crystals (Fig. 2b) consisting of the RIG-I SF2 ATPase domain missing two internal loops (Fig. 2c).

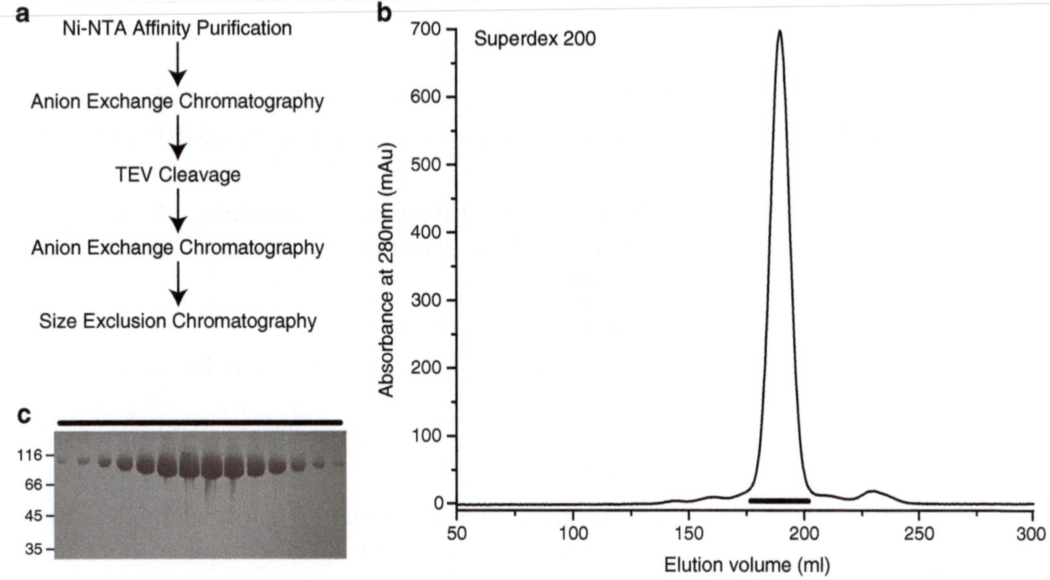

Fig. 1 Purification of mouse RIG-I 1-797. (**a**) Workflow of recombinant mouse RIG-I 1-797 purification. (**b**) Size exclusion chromatogram of RIG-I 1-797. Construct elutes in a single symmetric peak from HiLoad Superdex 200 26 60 (GEHealthcare) column. (**c**) SDS-PAGE analysis of size exclusion chromatography fractions. *Black horizontal line* in Fig. 1b corresponds to the *black line* over the gel photo

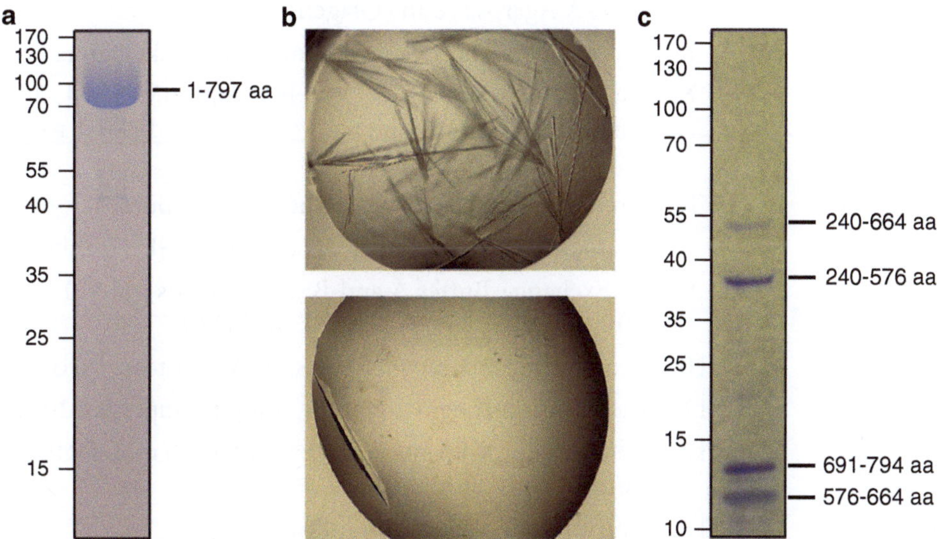

Fig. 2 RIG-I SF2 domain crystals. (**a**) SDS-PAGE analysis of recombinant RIG-I 1-797 before crystallization setup. (**b**) Crystal images of RIG-I SF2 domain. Different amount of crystals may form independent of protein or precipitant concentration. Thus, several drops should be set to obtain diffraction quality crystals. (**c**) SDS-PAGE analysis of RIG-I SF2 domain crystals. The crystals were collected and washed three times with reservoir solution and loaded on SDS-PAGE. The bands were assigned based on the ordered parts in the crystal structure (PDB Accession code: 3TBK)

2 Materials

2.1 Expression of Mouse RIG-I 1-797

1. Competent *E. coli* Rosetta (DE3) (Novagen) cells.
2. Modified pET vector expressing 6×His-MBP-TEV-RIG-I 1-797.
3. Luria Broth (LB) and LB agar plate containing 30 μg chloramphenicol (for pRARE selection), 100 μg/ml ampicillin (for pET selection).
4. Sterile flasks to grow bacteria.
5. A temperature-controlled shaker that can be set to 37 and 18 °C.
6. A spectrophotometer and cuvettes that can measure absorbance at 600 nm.
7. 0.5 M Isopropyl β-D-1-thiogalactopyranoside (IPTG), 200 μl of 0.5 M stock is required per 1 l of culture to obtain final concentration of 0.1 mM.

2.2 Purification of Mouse RIG-I 1-797

1. A high-speed centrifuge.
2. Lysis Buffer: 50 mM Tris–HCl pH 8.0, 0.5 M NaCl, 10 mM Imidazole pH 8.0, 10 mM β-mercaptoethanol.
3. Sonifier (Branson Sonifier 250 (Emerson Industrial Automation)).
4. Ni-NTA Agarose resin (Qiagen).
5. Disposable polypropylene columns with disks/filters.
6. Elution Buffer: 50 mM Tris–HCl, 100 mM NaCl, 200 mM Imidazole, 10 mM β-mercaptoethanol (adjusted to pH 9.0 at 4 °C).
7. Chromatography system with at least two pumps.
8. Anion exchange column (HiTrap Q HP (GE Healthcare)).
9. Ion Exchange Buffer A and B: 30 mM Tris pH 9.0 (at 4 °C), 100 mM (A)/1 M (B) NaCl, 1 mM DTT.
10. Recombinant tobacco etch virus (TEV) Protease (Tropea et al.)
11. Size exclusion chromatography column (Superdex 200).
12. Size Exclusion Chromatography Buffer: 20 mM TRIS pH 7.5, 150 mM NaCl, 1 mM DTT.
13. Centrifugal concentration units.
14. A spectrophotometer and cuvettes that can measure absorbance at 280 nm.

2.3 Crystallization of RIG-I 1-797 by In Situ Proteolysis

1. 50 mM 5′-(β,γ-imido)triphosphate (AMPPNP) in 100 mM Tris–HCl pH 7.5.
2. Subtilisin 1 mg/ml in 1 mM HCl and 2 mM $CaCl_2$.

3. Sitting drop crystallization plates and seals to cover them.
4. Reservoir Solution: 100 mM BIS-Tris pH 6.6 and 22 % (w/v) PEG3350. Prepare 1 M BIS-Tris pH 6.6 by dissolving 10.4 g in 40 ml H_2O, pH to 6.6 with HCl and complete volume to 50 ml with H_2O. Filter and store at room temperature. Prepare 50 % (w/v) PEG3350 by dissolving 25 g of PEG3350 in 50 ml H_2O. Filter and store at room temperature. For 1 ml of reservoir solution, add 100 μl of 1 M BIS-Tris pH 6.6, 440 μl of 50 % (w/v) PEG3350, and 460 μl filtered H_2O and mix thoroughly.

3 Methods

Carry out all procedures on ice or at 4 °C unless otherwise specified.

3.1 Expression of Mouse RIG-I 1-797

1. Transform competent Rosetta (DE3) *E. coli* with pET vector expressing 6His-MBP-TEV-RIG-I 1-797 and spread to LB agar plate containing 30 μg chloramphenicol, 100 μg/ml ampicillin. Incubate the plates overnight at 37 °C.
2. Remove the plate from the incubator in the morning and store at room temperature until late afternoon.
3. Pick up 4–5 colonies and inoculate a starter culture of 10 ml in LB containing 30 μg chloramphenicol, 100 μg/ml ampicillin per a liter of induction culture. Incubate overnight at 37 °C with 220 rpm) shaking.
4. Dilute 10 ml starter culture to 1 l LB containing 30 μg chloramphenicol, 100 μg/ml ampicillin (1:100 dilution of starter culture) and incubate at 37 °C with shaking until the OD_{600} reaches a value of 0.4–0.6.
5. Decrease the temperature of the incubator to 18 °C and wait until the culture cools down and reaches the OD_{600} of 0.6–0.8.
6. Induce the culture with adding 200 μl of 0.5 M (final 0.1 mM) IPTG and incubate overnight at 18 °C with 220 rpm shaking.

3.2 Purification of Mouse RIG-I 1-797

1. Harvest the bacteria expressing 6His-MBP-TEV-RIG-I 1-797 by centrifugation at 4 °C with 6,000×*g*. Discard the supernatant.
2. Resuspend the pellet in 50 ml lysis buffer by pipetting up and down and vortexing.
3. Sonicate (Branson Sonifier 250) the suspension with a 9.5 mm tip for a minute with 50 % duty cycle and 8–9 output control.

4. Centrifuge the lysate for 45 min at 16,000 × *g* at 4 °C.
5. Equilibrate 1 ml of Ni-NTA beads with the lysis buffer in a disposable polypropylene column.
6. Collect and apply the cleared lysate onto Ni-NTA and allow binding of the construct with gravity flow.
7. Wash unbound sample with 5–10 bead volume of lysis buffer.
8. Elute bound sample with 5 bead volume of elution buffer and collect in a tube.
9. Equilibrate HiTrap Q HP column with buffer A.
10. Filter the eluate and inject to the equilibrated HiTrap Q HP column and wash unbound sample.
11. Elute the bound sample with a gradient of 0–40 % of buffer B in 20 column volumes.
12. Pool fractions corresponding to the peak of the construct and measure the protein concentration at 280 nm, the correction value for 6His-MBP-TEV-RIG-I is 1.1. Add 1 mg of TEV protease per 30 mg of protein. Incubate overnight with gentle mixing at 4 °C.
13. Equilibrate HiTrap Q HP column with buffer A.
14. Adjust the pH of the sample to 9.0 and the NaCl concentration to 100 mM and filter.
15. Inject the sample to buffer A equilibrated HiTrap Q HP and elute the bound sample with a gradient of 0–40 % buffer B in 20 column volumes.
16. Pool the fractions corresponding to peak of RIG-I 1-797 (Peak 2) and concentrate down to the appropriate volume for the size exclusion chromatography (SEC) column. (Peak 1 corresponds to 6His-MBP.)
17. Equilibrate the SEC column with SEC buffer.
18. Inject the sample to SEC column and collect the elution fractions.
19. Equilibrate the centrifugal concentration unit with the SEC buffer.
20. Pool the fractions corresponding to peak of RIG-I 1-797 and apply onto the concentration unit.
21. Concentrate the sample up to 25–30 mg/ml by centrifugation as suggested by the manufacturer of the concentration unit.
22. Collect the sample at 25–30 mg/ml in an eppendorf tube and centrifuge for 10 min at 16,000 × *g* at 4 °C.
23. Aliquot and freeze the sample in liquid nitrogen and store at −80 °C.

3.3 Crystallization of RIG-I 1-797 by In Situ Proteolysis

1. Prepare the reservoir solution from stocks freshly and use 0.5 ml per 24-well sitting drop vapor diffusion plate.
2. Add 1 volume of 50 mM adenosine 5′-(β,γ-imido)triphosphate (AMPPNP) to 10 volumes of RIG-I 1-797 (25–30 mg/ml) and 1 mg of subtilisin per 500 mg of RIG-I 1-797.
3. Immediately centrifuge the sample for 10 min at 16,000 × *g* at 4 °C.
4. Set up sitting drops consisting of 1 μl of reservoir and 1 μl of RIG-I 1-797: AMPPNP: subtilisin mix and seal the plate (Important to do the setups as soon as possible to prevent extensive protease cleavage).
5. Store the plate at 20 °C. The crystals grow within 1–2 days.

4 Notes

1. Crystallization requires mg amounts of homogenous protein samples, thus expression and purification are crucial steps of crystallography. In this study, we have chosen to work with mouse RIG-I as unlike human RIG-I it was well expressed and soluble in *E. coli* cells. To ensure the quality of the sample during purification, it is important to do SDS-PAGE analysis of the fractions after each step, especially after size exclusion chromatography (an example can be seen in Fig. 1c). Recombinant full-length mouse RIG-I with N-terminal 6His-MBP tag can also be purified to homogeneity with the same method. The overall workflow is depicted in Fig. 1a.
2. In situ proteolysis in crystallization has been reported to be successful in several cases and its general applicability was tested using a variety of proteases on proteins that had remained to be reluctant to crystallization at the time [30, 31]. Thus, the method has proven its strength for many polypeptide chains. However, our experience showed many variables to be effective on the process. The first is the choice of the construct. We did exactly the same setups with three different RIG-I constructs (full-length, CARDless, and RDless) all consisting the crystallized ATPase domain; however, crystals were observed only with the construct that lacked the C-terminal regulatory domain. The second variable is the engine of in situ proteolysis, i.e., the protease. For this study, we tested trypsin, chymotrypsin, and subtilisin in same conditions and subtilisin was the only protease which yielded crystals. This phenomenon can easily be explained by its ability to attack any peptide bond independent of the primary structure unlike trypsin and chymotrypsin. However, in some cases selective cleavage activity of the protease might be necessary. The other variables that

should be taken into account while applying in situ proteolysis include the relative protease concentration in addition to the usual crystallization variables, such as choice of polypeptide concentration, screen, temperature, and time.

3. The analysis of the crystals on SDS-PAGE combined with the X-ray crystallographic structure obtained from these crystals suggested that subtilisin cleaved between CARDs and SF2 ATPase domain and removed two internal loops of SF2 resulting in a module that comprises SF2 domain (Fig. 2c).

Acknowledgments

This work was funded by National Institutes of Health grant U19AI083025 and grants from the Deutsche Forschungsgemeinschaft (DFG HO2489/3 and SFB455) and financial support from Center for Integrated Protein Science Munich to KPH.

References

1. Takeuchi O, Akira S (2010) Pattern recognition receptors and inflammation. Cell 140(6): 805–820
2. Yoneyama M et al (2004) The RNA helicase RIG-I has an essential function in double-stranded RNA-induced innate antiviral responses. Nat Immunol 5(7):730–737
3. Kang DC et al (2002) mda-5: an interferon-inducible putative RNA helicase with double-stranded RNA-dependent ATPase activity and melanoma growth-suppressive properties. Proc Natl Acad Sci U S A 99(2):637–642
4. Kovacsovics M et al (2002) Overexpression of Helicard, a CARD-containing helicase cleaved during apoptosis, accelerates DNA degradation. Curr Biol 12(10):838–843
5. Rothenfusser S et al (2005) The RNA helicase Lgp2 inhibits TLR-independent sensing of viral replication by retinoic acid-inducible gene-I. J Immunol 175(8):5260–5268
6. Yoneyama M, Fujita T (2008) Structural mechanism of RNA recognition by the RIG-I-like receptors. Immunity 29(2):178–181
7. Saito T et al (2007) Regulation of innate antiviral defenses through a shared repressor domain in RIG-I and LGP2. Proc Natl Acad Sci U S A 104(2):582–587
8. Yoneyama M et al (2005) Shared and unique functions of the DExD/H-box helicases RIG-I, MDA5, and LGP2 in antiviral innate immunity. J Immunol 175(5):2851–2858
9. Li X et al (2009) The RIG-I-like receptor LGP2 recognizes the termini of double-stranded RNA. J Biol Chem 284(20):13881–13891
10. Li X et al (2009) Structural basis of double-stranded RNA recognition by the RIG-I like receptor MDA5. Arch Biochem Biophys 488:23–33
11. Lu C et al (2010) Crystal structure of RIG-I C-terminal domain bound to blunt-ended double-strand RNA without 5′ triphosphate. Nucleic Acids Res 39:1565–1575
12. Wang Y et al (2010) Structural and functional insights into 5′-ppp RNA pattern recognition by the innate immune receptor RIG-I. Nat Struct Mol Biol 17:781–787
13. Hornung V et al (2006) 5′-Triphosphate RNA is the ligand for RIG-I. Science 314(5801): 994–997
14. Pichlmair A et al (2006) RIG-I-mediated antiviral responses to single-stranded RNA bearing 5′-phosphates. Science 314(5801):997–1001
15. Schlee M et al (2009) Recognition of 5′ triphosphate by RIG-I helicase requires short blunt double-stranded RNA as contained in panhandle of negative-strand virus. Immunity 31:25–34
16. Cui S et al (2008) The C-terminal regulatory domain is the RNA 5′-triphosphate sensor of RIG-I. Mol Cell 29(2):169–179
17. Lu C et al (2010) The structural basis of 5′ triphosphate double-stranded RNA recognition

by RIG-I C-terminal domain. Structure 18(8): 1032–1043

18. Kawai T et al (2005) IPS-1, an adaptor triggering RIG-I- and Mda5-mediated type I interferon induction. Nat Immunol 6(10):981–988
19. Xu LG et al (2005) VISA is an adapter protein required for virus-triggered IFN-beta signaling. Mol Cell 19(6):727–740
20. Seth RB et al (2005) Identification and characterization of MAVS, a mitochondrial antiviral signaling protein that activates NF-kappaB and IRF 3. Cell 122(5):669–682
21. Meylan E et al (2005) Cardif is an adaptor protein in the RIG-I antiviral pathway and is targeted by hepatitis C virus. Nature 437(7062): 1167–1172
22. Gack MU et al (2007) TRIM25 RING-finger E3 ubiquitin ligase is essential for RIG-I-mediated antiviral activity. Nature 446(7138): 916–920
23. Zeng W et al (2010) Reconstitution of the RIG-I pathway reveals a signaling role of unanchored polyubiquitin chains in innate immunity. Cell 141(2):315–330
24. Bamming D, Horvath CM (2009) Regulation of signal transduction by enzymatically inactive antiviral RNA helicase proteins MDA5, RIG-I, and LGP2. J Biol Chem 284(15):9700–9712
25. Civril F et al (2011) The RIG-I ATPase domain structure reveals insights into ATP-dependent antiviral signalling. EMBO Rep 12(11):1127–1134
26. Jiang F et al (2011) Structural basis of RNA recognition and activation by innate immune receptor RIG-I. Nature 479(7373):423–427
27. Kowalinski E et al (2011) Structural basis for the activation of innate immune pattern-recognition receptor RIG-I by viral RNA. Cell 147(2):423–435
28. Lu C et al (2011) Crystal structure of RIG-I C-terminal domain bound to blunt-ended double-strand RNA without 5′ triphosphate. Nucleic Acids Res 39(4):1565–1575
29. Luo D et al (2011) Structural insights into RNA recognition by RIG-I. Cell 147(2): 409–422
30. Dong A et al (2007) In situ proteolysis for protein crystallization and structure determination. Nat Methods 4(12):1019–1021
31. Wernimont A, Edwards A (2009) In situ proteolysis to generate crystals for structure determination: an update. PLoS One 4(4):e5094

Chapter 4

Isolation of RIG-I-Associated RNAs from Virus-Infected Cells

Andreas Schmidt, Andreas Linder, Nicolas Linder, and Simon Rothenfusser

Abstract

When a novel innate pattern recognition receptor (PRR) is identified, a question comes up immediately: Which molecular pattern(s) can it recognize? One approach that can be taken to answer this question for nucleic acid-binding receptors is the detailed analysis of synthetic ligands (DNA, RNA, or hybrids) to narrow in on the minimal patterns that activate a given receptor. However, this may not always lead to a satisfying answer. A complementary albeit technically more demanding way to tackle this question is to examine which nucleic acids are actually bound by the receptor in a setting of cellular infection. Here, we describe a basic protocol to isolate RNAs bound to the RNA receptors of the RIG-I-like helicase family from virus-infected cells via immunoprecipitation (IP). The isolated RNA can then be used to analyze its origin, characteristics, and immunostimulatory properties with a variety of methods.

Key words RNA, Immunoprecipitation, Virus, FLAG-tag, RIG-I-like helicases

1 Introduction

Infection of cultured primary cells or cell lines with viruses of different origin is one of the models to analyze the function of cytoplasmic PRRs like those of the RIG-I-like helicase (RLH) family. The minimal structural requirements for RNAs detected by RIG-I are already quite well defined [1, 2]. Here, the focus of research has shifted to identifying the nature of the ligands that activate RIG-I during viral infection in the cytoplasm [3, 4]. Independently of the method employed to purify putative ligands, a good hypothesis about possible candidates helps to choose the appropriate downstream applications to analyze the bound ligands and identify them. For RIG-I and the viruses typically recognized by this receptor such a framework of likely ligands exists [5]. In other cases generating a good working hypothesis of how possible ligands

Hans-Joachim Anders and Adriana Migliorini (eds.), *Innate DNA and RNA Recognition*, Methods in Molecular Biology, vol. 1169, DOI 10.1007/978-1-4939-0882-0_4, © Springer Science+Business Media New York 2014

could look like may be less straightforward. This does not preclude their identification altogether, but may complicate the validation process.

RIG-I recognizes segmented and non-segmented positive- or negative-strand RNA viruses and participates in the recognition of double-stranded RNA viruses [6]. We use vesicular stomatitis virus (VSV) from the order *Mononegavirales* as a model system. It replicates in the cytoplasm of infected cells and produces a range of RNAs during its replication cycle some of which bear the structural features that are necessary for recognition by RIG-I. These RNAs include but may not be limited to the viral genomic and anti-genomic RNA and the leader and trailer transcripts [7].

The method presented here is aimed at purifying RIG-I-bound RNAs from VSV-infected cells in an unbiased fashion for later analysis for example by next-generation-sequencing (NGS) approaches. It is based on IP of the receptor from lysates of infected cells and subsequent isolation of bound RNA. In principle, this approach is not limited to RIG-I itself, but is applicable to the other members of the RLR-family of receptors, MDA-5 and Lgp2, or potential RNA receptors of similar function.

We describe here a protocol in which we precipitate the receptor from stable cell lines that express a FLAG-tagged version of the protein under the control of a tetracycline-inducible promoter. The RNA is isolated by a method that is able to yield RNA of vastly different lengths including the genomic RNA in the range of 10 kb as well as small noncoding RNAs that are only several bases long. The advantage of using tagged proteins is the availability of well-characterized commercial antibodies, in this case even covalently coupled agarose beads. Also, peptides are available to elute the protein–RNA complex from the beads after IP increasing the specificity of the RNA recovery process.

Using a stable cell line guaranties a rather uniform medium level expression of the protein of interest in the cell population to be analyzed making the isolation process reproducible. An obvious problem is the location of the epitope tag. In this case we use an amino-terminal triple FLAG-tag fused to the caspase-activation and recruitment domains (CARDs) of RIG-I. We have used functional assays to confirm activity of the overexpressed protein.

Because protein–RNA interaction assays seem to be more prone to unspecific binding than protein–protein interaction assays, care must be taken to use meaningful controls. We use several controls including unspecific antibodies and cell lines that do not express the protein of interest. Here, we include cells that do not carry the inducible RIG-I-Flag expression cassette at all (Hek293 FlpIn cells, *see* Subheading 3), because we have noted that in our hands leaky expression from the non-induced promoter can lead to significant pull-down of FLAG-tagged RIG-I and consequently RIG-I-bound RNA. Another important consideration is the amount of starting material that is to be used to generate sufficient amount of eluted RNA for downstream applications. We typically start with 40 million cells to generate 2–3 μg of RNA, which is sufficient for downstream analysis by qRT-PCR, biological activity assays, and NGS.

2 Materials

2.1 Reagents, Buffers and Media

1. Medium I: Dulbecco's DMEM high glucose medium, 10 μg/ml ciprofloxacin, 100 μg/ml hygromycin,15 μg/ml blasticidin, 10 % v/v FCS.
2. Medium II: Dulbecco's DMEM high glucose medium, 10 μg/ml ciprofloxacin, 100 μg/ml zeocin, 15 μg/ml blasticidin, 10 % v/v FCS.
3. Opti-MEM (Invitrogen).
4. 1× PBS, sterile.
5. Lysis buffer: 20 mM Tris–HCl (pH 7.5), 150 mM NaCl, 0.25 % NP-40, 1.5 mM MgCl2, 1 mM NaF, 400 U/ml RNase inhibitor, protease inhibitor cocktail.
6. 2× Laemmli Buffer: 65.8 mM Tris–HCl, pH 6.8, 2.1 % SDS, 26.3 % (w/v) glycerol, 0.01 % bromophenol blue.
7. RNase ZAP (Invitrogen).
8. Preclearing beads: Sepharose CL-4B.
9. FLAG beads: ANTI-FLAG M2 affinity gel.
10. IgG beads: Mouse IgG–agarose saline suspension.
11. Tetracycline.
12. FLAG-RIG-I FlpIn cells.
13. FlpIn cells.
14. TBS (50 mM Tris, 150 mM NaCl, pH 7.6).
15. 3×FLAG peptide stock solution in 20 μl aliquots at 5 mg/μl.
16. 3×FLAG peptide working solution 100 μg/ml (20 μl stock in 1 ml TBS).
17. Spin column: Micro Bio-Spin chromatography columns (Bio-Rad).
18. miRNeasy Micro Kit.
19. 1205 Lu melanoma cells.
20. Lipofectamine RNAiMax reagent (Invitrogen).
21. Human IP-10 ELISA kit.
22. Vesicular stomatitis virus (Indiana strain).

3 Methods

3.1 Purification of RIG-I-Bound RNAs from VSV-Infected Cells

Hek293 FLAG-RIG-I cells were cultivated in medium I. For the control condition lacking the over-expressed protein Hek293 FlpIn cells were cultivated in medium II. We typically run the conditions:

- Virus-infected and FLAG-RIG-I-induced cells immunoprecipitated with anti-FLAG-coupled agarose.
- Virus-infected and FLAG-RIG-I-induced cells immunoprecipitated with IgG-coupled agarose.
- Virus-infected cells without FLAG-RIG-I expression cassette immunoprecipitated with anti-FLAG-coupled agarose.
- Non-infected FLAG-RIG-I-induced cells immunoprecipitated with anti-FLAG-coupled agarose.

1. For every condition (see Fig. 1), plate 40 million cells in four 175 cm^2 cell culture flasks with 10 million cells per flask in 15 ml of DMEM medium (*see* **Note 1**).
2. After incubation overnight, 24 h prior to infection, treat cells with 1 μg/ml of tetracycline to induce expression of the target protein.
3. Shortly before virus infection count the cells of your reference flask (*see* **Note 1**) and prepare virus-containing medium using 15 ml Opti-MEM per flask at a multiplicity of infection (MOI) of 1. After 24 h carefully remove medium and substitute with virus containing medium. Allow infection to occur for 9 h at 37 °C.
4. After 9 h most cells will have detached from the bottom of the flask. Shake off the rest of the cells until all cells have become detached (*see* **Note 2**). Wash flasks with 10 ml of ice-cold PBS (*see* **Note 3**). Collect cells, media and PBS in 50 ml tubes and put them on ice.
5. Centrifuge cells at 400 × *g* for 7 min at 4 °C and discard the supernatant. Resuspend the cell pellet in 1 ml of PBS and pool cells from different tubes of one condition and add 30 ml cold PBS. Centrifuge again and repeat this washing procedure three times.
6. After washing resuspend cells in 1 ml of lysis buffer (*see* **Note 4**). Transfer lysates into 1.5 ml Eppendorf vials and homogenize with a 1 ml syringe and 20 G needle.
7. Centrifuge lysates at 4 °C for 15 min at 10,000 × *g* and collect the supernatants.
8. In the meantime prepare uncoated sepharose beads for pre-clearing of the lysates by washing 500 μl slurry per condition two times in 1 ml Lysis buffer in 1.5 ml Eppendorf tubes (*see* **Note 5**).

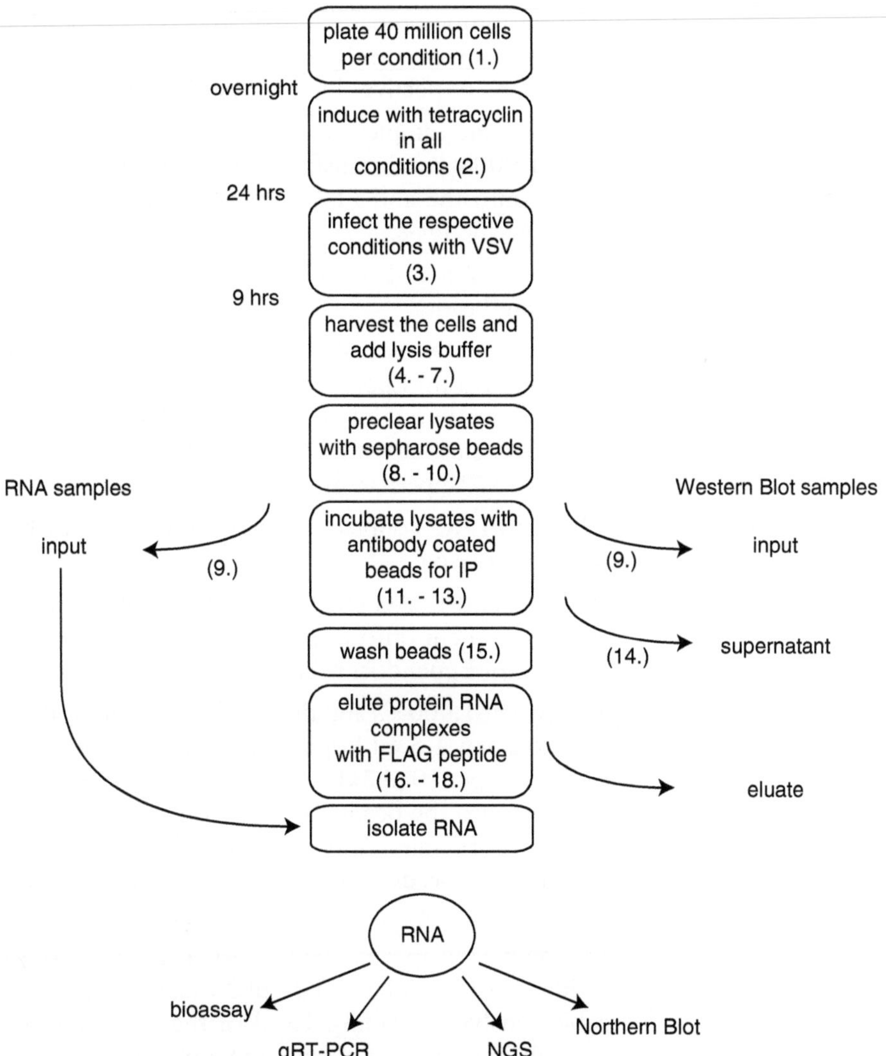

Fig. 1 Flowchart of the immunoprecipitation protocol. The flowchart depicts the important steps in the IP protocol from growing the cells, lysis, to IP and finally analysis of the eluate (step numbers are indicated in *brackets*). Steps where part of the sample are removed for subsequent analysis and the route for processing them are indicated by *arrows*

9. Take 100 μl of the lysates for RNA isolation and freeze at −80 °C. Take 20 μl for western blot analysis. Dilute the western blot sample in 2× Laemmli Buffer and freeze at −20 °C. These samples will be referred to as "input".
10. Add the remaining 880 μl of the lysates to the pre clearing beads prepared in **step 8**. Seal the Eppendorf tubes with Parafilm and put them into a 50 ml Falcon tube. Incubate for 1 h on a tube roller at 4 °C.

11. In the meantime prepare anti FLAG-coupled beads or beads coated with unspecific IgGs for each condition, by washing 400 μl of the respective slurry in lysis buffer twice.
12. Spin down the samples containing the pre clearing beads at 4 °C at 1,000 × *g* and transfer the supernatant of the pre clearing step onto the anti-FLAG coupled beads or beads coated with unspecific IgGs.
13. Incubate for 2 h at 4 °C on a tube roller as described for the pre clearing beads.
14. Spin down the beads at 4 °C at 1,000 × *g*. Transfer the supernatant into 1.5 ml Eppendorf tubes. Take 20 μl of the supernatant for western blot analysis and dilute in 2× Laemmli buffer. Freeze at −20 °C. These samples will from now on be referred to as "supernatant".
15. Wash anti-FLAG coupled beads in 1 ml of lysis buffer five times (*see* **Note 6**).
16. Resuspend anti-FLAG coupled beads in 200 μl of lysis buffer and transfer them onto a spin column. Put the spin columns into a 1.5 ml Eppendorf tube.
17. Spin down at 1,000 × *g* and discard flow-through. Put spin column into a new 1.5 ml Eppendorf tube. This tube will contain the eluate. Pipet 10 μl 3×FLAG peptide stock solution and 3 μl RiboLock onto the beads. Incubate on a shaker at room temperature for 10 min. Add another 137 μl 3×FLAG working solution to the beads and incubate for 10 min (*see* **Note 7**).
18. Spin down at 3,000 × *g* for 3′ at 4 ° C to elute the bound protein–RNA complex in approximately 150 μl. Take 3 μl of the eluate and dilute in 17 μl TBS and 20 μl 2× Laemmli buffer for western blot analysis. Loading 15 μl of the western blot samples will result in 0.75 % of every sample per lane. The rest of the eluate will be used for RNA extraction. These samples will from now on be referred to as "eluate".

3.2 RNA Isolation

For RNA isolation we use the QIAGEN miRNeasy Mini Kit. This kit is designed to purify relatively short microRNAs along with longer RNAs. This would allow us to isolate RNA species of a large range of sizes as they are typically produced by viruses. Classical phenol–chloroform extraction did not work very well in our hands.

If using the QIAGEN miRNeasy Mini Kit, add 700 μl of QIAZOL lysis reagent to the 100 μl of "input", "supernatant" and "eluate". Homogenize the samples by pipetting. The rest of the RNA isolation can be performed according to manufacturer's instructions. We asses the quantity and quality of the RNA preparation by measuring a UV-absorption spectrum in a NanoDrop

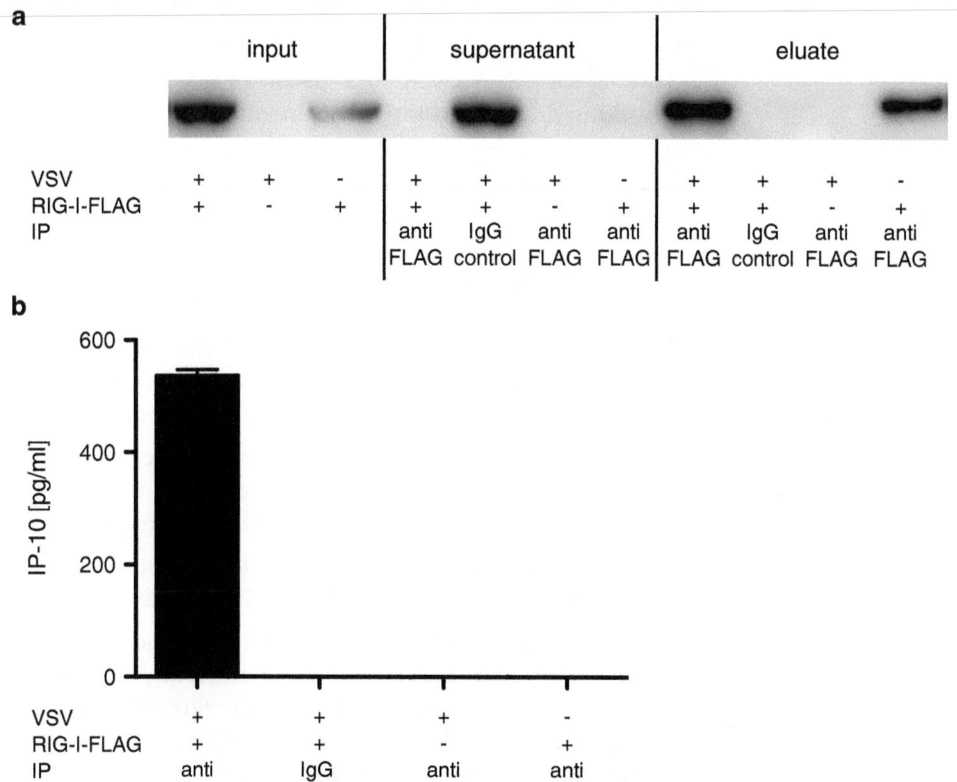

Fig. 2 Analyses to validate a successful IP experiment. (**a**) Immunoblot showing the analysis of samples from different steps of the IP. Samples of the lysates from VSV-infected (+) or non-infected (–) cells, induced (+) or not induced (–, i.e., Hek293 FlpIn cells) for FLAG-tagged RIG-I (input, step 9), of the respective supernatants after FLAG or control IP (supernatant, step 14), and of the eluates after treatment of all beads with FLAG peptide (eluate, steps 16–18) where processed for SDS-PAGE and western blotting. The membrane was stained with anti-FLAG antibody and each lane represents 0.75 % of the input, supernatant or eluate for every IP condition. Note that conditions which differ only in IP-beads (FLAG and IgG) share a common input. (**b**) To test the immunostimulatory activity of the isolated RNAs 1250 Lu melanoma cells were transfected with 100 ng of RNA obtained from the different IP conditions using Lipofectamine RNAiMax as a transfection reagent. Production of the interferon-induced cytokine IP-10 was measured after 24 h in the culture supernatants of these cells via ELISA

photometer. We usually end up with 50–100 ng/μl of RNA in a volume of 50 μl. In order to control the efficiency of the IP perform a western blot, staining for the FLAG-tag using an anti FLAG antibody (Fig. 2a).

3.3 Bioassay

In order to determine whether immunostimulatory RLR-ligands have been pulled down, we transfect the isolated RNA into 1205 Lu cells and measure IP-10 as a marker of RLR-pathway activation in the supernatant after 24 h via enzyme-linked immunosorbent assay (ELISA) (Fig. 2b). Alternatively an IFN-promoter reporter cell line can be used. We typically plate 8,000 1205 Lu cells in 100 μl medium in a 96-well dish. On the next day we transfect 10–100 ng of RNA using 0.2 μl of Lipofectamine RNAiMax reagent diluted in 50 μl Opti-MEM medium.

4 Notes

1. In order to determine the exact number of cells on the day of infection and in order to account for differences in proliferation, we recommend plating an additional flask for every cell line and counting these cells prior to infection. This allows adapting multiplicity of infection (MOI) more accurately.
2. Special care must be taken with the cells of the uninfected control. For Hek293 we have found the use of trypsin unnecessary, but it might be necessary for other cell types.
3. From this step on the samples should be kept on ice to minimize degradation of the RNA–protein complexes.
4. An RNase-free working environment is crucial for all steps from here on. We clean all surfaces with RNaseZAP. As another important measure we use RNase-free filter tips to pipet our samples. Where appropriate we use RNase inhibitors in our buffers (see description in the protocol).
5. When handling beads we find it advantageous to cut the filter tips in order to allow easier pipetting.
6. We have noted that it is important to wash extensively in order to minimize unspecific binding. In addition a selection of adequate negative controls is crucial for correct interpretation of the data.
7. We have found the elution to be more efficient if we add the concentrated FLAG peptide solution first and add TBS in a second step as opposed to preparing a prediluted elution buffer.

References

1. Schmidt A et al (2009) 5′-Triphosphate RNA requires base-paired structures to activate antiviral signaling via RIG-I. Proc Natl Acad Sci U S A 106(29):12067–12072
2. Schlee M et al (2009) Recognition of 5′ triphosphate by RIG-I helicase requires short blunt double-stranded RNA as contained in panhandle of negative-strand virus. Immunity 31(1):25–34
3. Rehwinkel J et al (2010) RIG-I detects viral genomic RNA during negative-strand RNA virus infection. Cell 140(3):397–408
4. Baum A, Sachidanandam R, Garcia-Sastre A (2010) Preference of RIG-I for short viral RNA molecules in infected cells revealed by next-generation sequencing. Proc Natl Acad Sci U S A 107(37):16303–16308
5. Gerlier D, Lyles DS (2011) Interplay between innate immunity and negative-strand RNA viruses: towards a rational model. Microbiol Mol Biol Rev 75(3):468–490
6. Ramos HJ, Gale M Jr (2011) RIG-I like receptors and their signaling crosstalk in the regulation of antiviral immunity. Curr Opin Virol 1(3):167–176
7. Conzelmann KK (1998) Nonsegmented negative-strand RNA viruses: genetics and manipulation of viral genomes. Annu Rev Genet 32:123–162

Chapter 5

Structure Modeling of Toll-Like Receptors

Jing Gong and Tiandi Wei

Abstract

Toll-like receptors (TLRs) recognize invasion of microbial pathogens and initiate innate immune responses that are essential for inhibiting pathogen dissemination and for the development of acquired immunity. To understand how these receptors work, it is crucial to investigate them from a structural perspective. High-throughput genome sequencing projects have led to the identification of more than 3,000 TLR sequences. However, only several structures of TLRs have been determined because structure determination by X-ray diffraction or nuclear magnetic resonance spectroscopy experiments remains difficult and time-consuming. Protein structure modeling methods are powerful tools for bridging the gap between sequence determination and structure determination. Due to different repeat numbers and distinct arrangements of leucine-rich repeats (LRRs) contained in TLR ectodomains, an automated homology modeling method often failed to predict a proper model. Here, we describe an LRR template assembly method for homology modeling of TLRs. This method was successfully validated through the comparison of a predicted model with the crystal structures, and showed better performance than other Protein structure modeling tools. The resulting models can be used to perform protein–ligand interaction studies or to design mutagenesis experiments, and hence to investigate TLR ligand-binding mechanisms.

Key words Homology modeling, Template assembly, Leucine-rich repeat, TLR structure

1 Introduction

Toll-like receptors (TLRs) are a group of pattern-recognition receptors (PRR) that are expressed by cells of the innate immune system, such as macrophages and dendritic cells, and specifically recognize pathogen-associated molecular patterns (PAMPs) within microbes. The progress of genome sequencing projects has led so far to the identification of 13 subgroups of TLRs in mammalian genomes, 10 in humans and 13 in mice [1], and more than 20 in non-mammalian genomes [2]. All TLRs have a common domain organization, with an extracellular ectodomain and an intracellular Toll/IL-1 receptor homology (TIR) domain [3] (Fig. 1). Both domains are joined by a single transmembrane helix stretch (ca. 20 amino acids), which determines the subcellular localization of TLRs [4]. The ectodomain is responsible for the recognition of

Hans-Joachim Anders and Adriana Migliorini (eds.), *Innate DNA and RNA Recognition*, Methods in Molecular Biology, vol. 1169, DOI 10.1007/978-1-4939-0882-0_5, © Springer Science+Business Media New York 2014

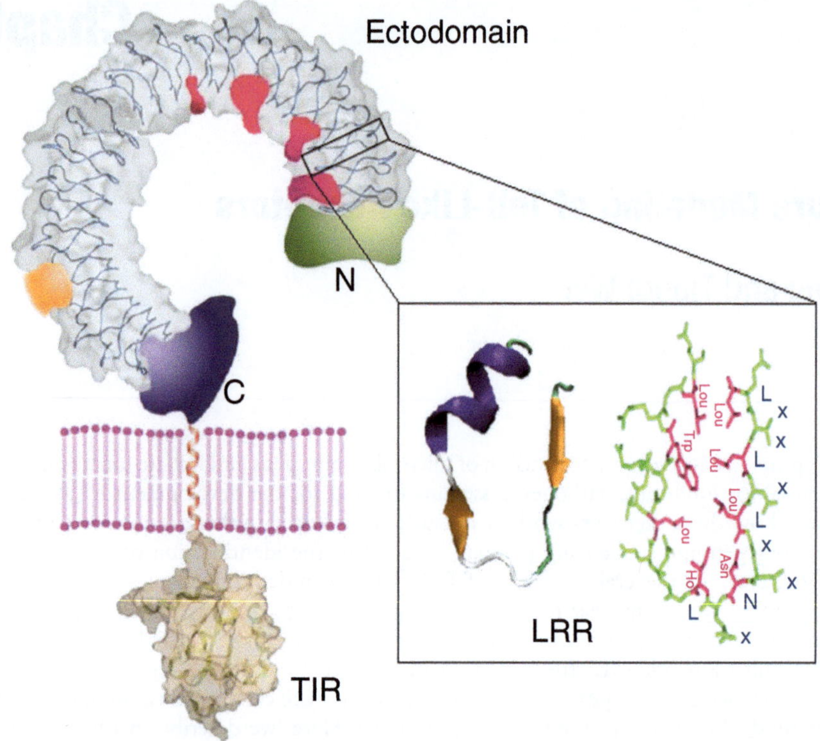

Fig. 1 Structure of the TLR and LRR

common structural patterns in various microbial molecules, such as lipopolysaccharides, lipopeptides, nucleotides, or bacterial flagellins [5, 6]. The TIR domains of TLRs are associated with the intracellular signaling cascade leading to the nuclear translocation of the transcription factor NF-κB [7].

A TLR ectodomain contains 19–27 consecutive leucine-rich repeat (LRR) motifs with two terminal LRR modules (LRRNT and LRRCT) at both ends [2] (Fig. 1). LRRs exist in more than 30,000 known proteins, and more than 160 crystal structures of these proteins have been deposited in the Protein Data Bank (PDB). In every case, the protein adopts an arc (less LRRs) or horseshoe (more LRRs) shape. An individual LRR motif is defined as an array of 20–30 amino acids that is rich in the hydrophobic amino acid leucine. All LRR sequences can be divided into a conserved segment and a variable segment. The conserved segments, with consensus sequence LxxLxLxxNxL, generate the concave surface of the LRR arc or horseshoe by forming parallel β-strands, while the variable parts form its convex surface, consisting of helices or loops. The terminal LRRNT and LRRCT modules stabilize the protein structure by shielding its hydrophobic core from exposure to solvent.

To date, the crystal structures of the ectodomains of human TLR1, 2, 3, and 4, mouse TLR2, 3, 4, and 6, and zebrafish TLR5, and the TIR domains of human 1, 2, and 10 have been determined [8–18]. High-throughput genome sequencing projects, however, have led to the identification of more than 3,000 TLR sequences. Thus, the structures of most TLRs are still unknown because structure determination by X-ray diffraction or nuclear magnetic resonance spectroscopy experiments remains difficult and time-consuming. Protein structure modeling methods are powerful tools for bridging the gap between sequence determination and structure determination.

Homology modeling, also referred to as comparative modeling, is currently the most accurate computational method for protein structure prediction. This approach constructs a three-dimensional model for a target protein sequence from a three-dimensional template structure of a homologous protein. Therefore, the quality of the homology model strongly depends on the sequence identity between the target and template. Below 30 % identity, serious errors may occur [19]. As TIR domain is highly conserved in structure across all TLRs and the downstream signaling adaptor molecules, the known crystal structures of TIR domains provide excellent templates for the modeling of structure-unknown TIR domains. Therefore, the three-dimensional model of a TIR domain can be just created by fully automated homology modeling software tools, such as SWISS-MODEL [20] and I-TASSER [21]. In the case of TLR ectodomain, however, due to different repeat numbers and distinct arrangements of LRRs in the ectodomains, a suitable full-length template with a sufficiently high sequence identity to the target is often missing. Thus, the automated homology modeling tools fail to return a satisfactory result (*see* **Note 1**). This limitation can be overcome by assembling multiple LRR templates [22]. The present article provides a protocol for this assembly method (*see* **Note 2**). In this approach, the most similar (at the sequence level) structure-known single LRR is searched for as a local template for each LRR segment in the target sequence. Such an LRR template may be derived from TLRs or other LRR proteins. All local template sequences are then combined to generate a multiple sequence alignment with the full-length target sequence. In this way, a high-quality model can be created, even if no adequate full-length template is available (*see* **Note 3**).

2 Materials

2.1 Model Construction

1. TollML database: a database of sequence motifs of TLRs [2]. TollML includes all known TLR protein sequences extracted from the NCBI protein database [23]. Each sequence was semi-automatically partitioned into four parts: signal peptide,

ectodomain, transmembrane domain, and TIR domain. Each ectodomain was semi-automatically partitioned into individual LRRs.

2. LRRML database: a conformational LRR database [24]. LRRML archives individual LRR structures that were manually identified from all known LRR protein structures extracted from the Protein Data Bank [25].
3. Global sequence alignment using EMBOSS Needle [26].
4. Online program for protein structure modelling of TIR domain: I-TASSER [21].
5. Local program for protein structure modelling of ectodomain: MODELLER 9.11 [27].
6. Web server for automated modelling of loops in protein structures: ModLoop [28].
7. Online model quality assessment programs: ProQ [29], ModFold [30], and ProCheck [31].

2.2 Model Analysis

1. Three-dimensional molecular viewer: VMD [32].
2. Macromolecular electrostatics calculation program: APBS [33].
3. Protein superposition server: SuperPose [34].
4. Protein–ligand interaction prediction program: AutoDock [35].

3 Methods

3.1 Model Construction

1. The full-length amino-acid sequence of the target TLR is extracted from TollML (http://tollml.lrz.de/). The domain partition and LRR partition were already annotated by TollML.
2. The amino-acid sequence of the TIR domain is directly copied and pasted into the input field of the Web page of I-TASSER (http://zhanglab.ccmb.med.umich.edu/I-TASSER/). Five best resulting models will be returned 1 day after submission.
3. For each LRR sequence contained in the ectodomain, a three-dimensional LRR structure in .pdb file with the highest sequence identity is selected as a template from LRRML through a sequence similarity search tool implemented in LRRML (http://tollml.lrz.de/).
4. A multiple sequence alignment of the full-length ectodomain sequence with all its local LRR template sequences is generated using EMBOSS Needle (http://www.ebi.ac.uk/Tools/psa/emboss_needle/), with each template comprising one alignment line. For instance, the mouse TLR3 ectodomain has a total of 25 LRRs and accordingly has 25 templates. The associated multiple sequence alignment is then composed of 26 lines.

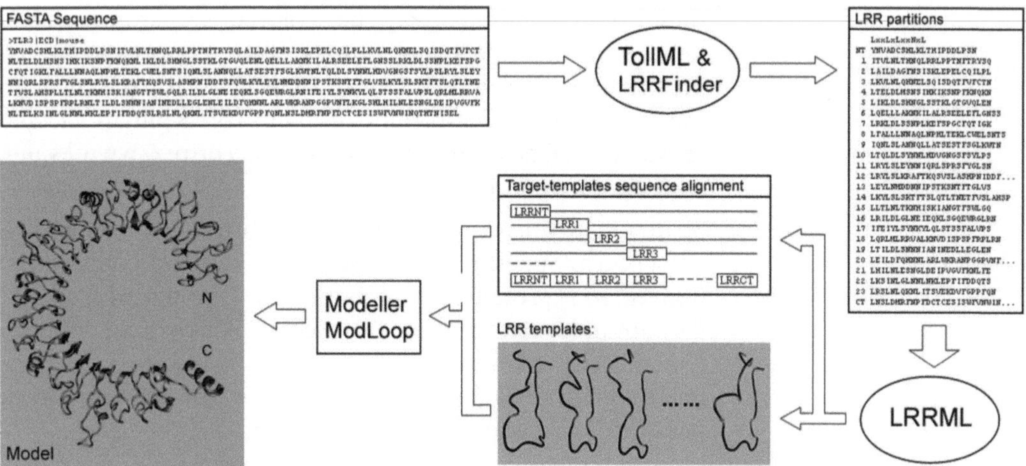

Fig. 2 Flowchart of the template assembly method for homology modeling of TLR ectodomains

5. The multiple alignment file and the corresponding template structures (.pdb files) are inputted into the locally installed program MODELLER (http://www.salilab.org/modeller/).
6. MODELLER calculates the initial three-dimensional coordinate file (.pdb file) for the model. A given number of models are returned. A schematic flowchart of the entire modeling procedure is shown in Fig. 2.
7. The resulting models are inputted to the online model quality assessment programs: ProQ (http://www.sbc.su.se/~bjornw/ProQ/ProQ.cgi), ModFold (http://www.reading.ac.uk/bioinf/ModFOLD/) and ProCheck (http://nihserver.mbi.ucla.edu/SAVES/), respectively. Model quality assessment involves analysis of geometry, stereochemistry, and energy distributions of the models. The best model is selected from the candidate models according to the assessment scores (*see* **Note 4**).
8. Sometimes the best model may have one or more local regions that are of low quality assessed by the model quality assessment programs. ModLoop (http://modbase.compbio.ucsf.edu/modloop/) is used to rebuild the coordinates of those regions to improve the quality of the model. The start and end positions of the regions to be rebuilt together with the model file (.pdb file) are inputted into ModLoop. A new model that is modified at the given regions is returned. This step is optional (*see* **Note 5**).

3.2 Model Analysis

1. Structure representation. VMD is a molecular visualization program for displaying, animating, and analyzing large biomolecular systems using three-dimensional graphics and built-in scripting. Molecules may be drawn as lines, bonds–spheres, ribbons, secondary structure cartoons, surfaces, etc. Each of

these representations may be colored by atom, residue, segment, molecule name, mass, charge, backbone, etc. Furthermore, a very useful function of VMD is to calculate the molecular surface electrostatics using its APBS plugin. Detailed assistance refers to the VMD tutorial (http://www.ks.uiuc.edu/Research/vmd/).

2. Structure comparison. SuperPose (http://wishart.biology.ualberta.ca/SuperPose/) is used to match and compare two homologous protein structural models. Two model files (.pdb file) are uploaded into the Web server. A new model file that contains the two superimposed models is returned. The root mean square deviations that measure the structural differences between both models are also reported.

3. Analysis of protein–ligand interactions. Docking is a computer simulation method that determines the preferred orientation of one molecule to a second when bound to each other to form a stable complex. AutoDock is a free downloadable automated docking tool. It calculates the most reasonable interacting models for two input models (.pdb file) based on the shape complementarity, electrostatic complementarity and hydrophobicity of molecular surfaces. Detailed assistance refers to the AutoDock tutorial (http://autodock.scripps.edu/).

4 Notes

1. TLR ectodomain and TIR domain have different structural organizations. The routine homology modeling software tools are suitable for the TIR domains but usually do not work for the ectodomains. In particular, the ectodomains of nucleic acid-specific TLRs, viz., TLR3, 7, 8, and 9, contain more LRRs than other TLRs; so there is no full-length template with a sufficiently high sequence identity. For comparison purposes, the mouse TLR3 ectodomain was modeled with a standard profile-profile alignment-aided full-length template recognition method in previous work [22]. The output model showed a serious structural disorder spanning LRR6-10, which were interwoven with one another (Fig. 3a). The LRR6-10 on the crystal structure, however, form a regular solenoid structure with an α-helix in LRR8 (Fig. 3c). By contrast, the model generated by the template assembly method (Fig. 3b) showed high structural similarity to the crystal structure. The template assembly method reveals its particular strength in situations where no adequate full-length templates are available. Nevertheless, the template assembly method is currently not an automated program. Several procedures still need manual check.

2. This template assembly approach can be extended to other repetitive proteins.

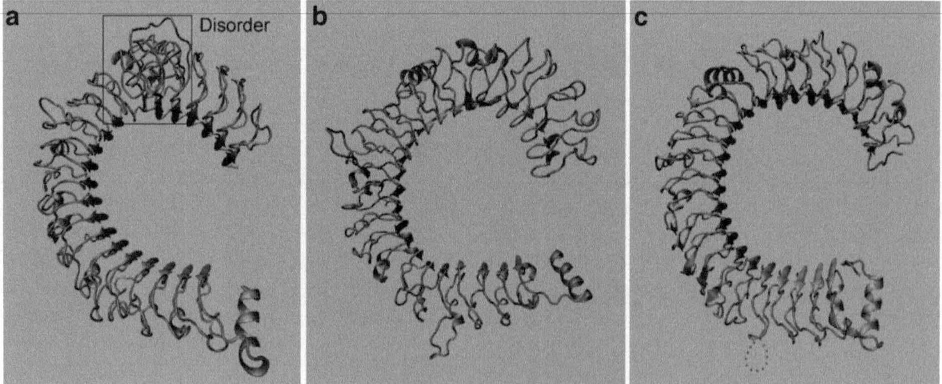

Fig. 3 Homology models and crystal structure of the mouse TLR3 ectodomain. (**a**) The homology model based on the standard method. The *framed region* exhibits serious disorder. (**b**) The homology model based on the template assembly method. (**c**) The crystal structure (PDB code: 3CIG). The *dotted region* is an insertion on LRR20 that is missing in the crystal structure

3. This approach basically relies on the TollML and LRRML databases. Timely data update of both databases can improve the accuracy of the modeling. We usually maintain the databases every 2 months. For very new TLR sequences that are not included in TollML yet, their LRR partitions can be done using the LRRFinder program on the TollML Web page (http://tollml.lrz.de:8081/exist/rest//db/tollml/finder/lrrfinder.xq) by inputting the sequence and clicking "submit."
4. During the model evaluation (Subheading 3.1, **step 7**) the initially selected model is sometimes found to be unsatisfactory. In this case, **steps 4–7** need to be repeated. The sequence alignment between target and templates is manually modified and new models are generated using MODELLER until acceptable evaluation results are obtained.
5. Some LRRs contain an insertion loop with length varying from 5 to 20 amino acids in their variable region. These loops are flexible in the in vivo environment, and often contribute to ligand-binding sites; therefore, they should be treated with extra care. In addition to rebuild the coordinates of the loops with ModLoop, the loops should be treated as flexible in ligand docking. If condition permits, molecular dynamics simulations can be performed to optimize the structures of loops.

Acknowledgments

This work was supported by the Independent Innovation Foundation of Shandong University, China (No. 2011HW009) and the Promotive Research Fund for Excellent Young and Middle-aged Scientists of Shandong Province, China (No. BS2012SW010).

References

1. Kumar H, Kawai T, Akira S (2009) Pathogen recognition in the innate immune response. Biochem J 420(1):1–16
2. Gong J, Wei T, Zhang N, Jamitzky F, Heckl WM, Rossle SC, Stark RW (2010) TollML: a database of Toll-like receptor structural motifs. J Mol Model 16(7):1283–1289
3. Brodsky I, Medzhitov R (2007) Two modes of ligand recognition by TLRs. Cell 130(6): 979–981
4. Barton GM, Kagan JC, Medzhitov R (2006) Intracellular localization of Toll-like receptor 9 prevents recognition of self DNA but facilitates access to viral DNA. Nat Immunol 7(1):49–56
5. West AP, Koblansky AA, Ghosh S (2006) Recognition and signaling by Toll-like receptors. Annu Rev Cell Dev Biol 22:409–437
6. Akira S, Hemmi H (2003) Recognition of pathogen-associated molecular patterns by TLR family. Immunol Lett 85(2):85–95
7. Leulier F, Lemaitre B (2008) Toll-like receptors: taking an evolutionary approach. Nat Rev Genet 9(3):165–178
8. Jin MS, Kim SE, Heo JY, Lee ME, Kim HM, Paik SG, Lee H, Lee JO (2007) Crystal structure of the TLR1-TLR2 heterodimer induced by binding of a tri-acylated lipopeptide. Cell 130(6):1071–1082
9. Choe J, Kelker MS, Wilson IA (2005) Crystal structure of human Toll-like receptor 3 (TLR3) ectodomain. Science 309(5734):581–585
10. Bell JK, Botos I, Hall PR, Askins J, Shiloach J, Segal DM, Davies DR (2005) The molecular structure of the Toll-like receptor 3 ligand-binding domain. Proc Natl Acad Sci U S A 102(31):10976–10980
11. Liu L, Botos I, Wang Y, Leonard JN, Shiloach J, Segal DM, Davies DR (2008) Structural basis of Toll-like receptor 3 signaling with double-stranded RNA. Science 320(5874): 379–381
12. Kim HM, Park BS, Kim JI, Kim SE, Lee J, Oh SC, Enkhbayar P, Matsushima N, Lee H, Yoo OJ et al (2007) Crystal structure of the TLR4-MD-2 complex with bound endotoxin antagonist Eritoran. Cell 130(5):906–917
13. Kang JY, Nan X, Jin MS, Youn SJ, Ryu YH, Mah S, Han SH, Lee H, Paik SG, Lee JO (2009) Recognition of lipopeptide patterns by Toll-like receptor 2-Toll-like receptor 6 heterodimer. Immunity 31(6):873–884
14. Ohto U, Fukase K, Miyake K, Shimizu T (2012) Structural basis of species-specific endotoxin sensing by innate immune receptor TLR4/MD-2. Proc Natl Acad Sci U S A 109(19):7421–7426
15. Yoon SI, Kurnasov O, Natarajan V, Hong M, Gudkov AV, Osterman AL, Wilson IA (2012) Structural basis of TLR5-flagellin recognition and signaling. Science 335(6070):859–864
16. Xu Y, Tao X, Shen B, Horng T, Medzhitov R, Manley JL, Tong L (2000) Structural basis for signal transduction by the Toll/interleukin-1 receptor domains. Nature 408(6808): 111–115
17. Tao X, Xu Y, Zheng Y, Beg AA, Tong L (2002) An extensively associated dimer in the structure of the C713S mutant of the TIR domain of human TLR2. Biochem Biophys Res Commun 299(2):216–221
18. Nyman T, Stenmark P, Flodin S, Johansson I, Hammarstrom M, Nordlund P (2008) The crystal structure of the human Toll-like receptor 10 cytoplasmic domain reveals a putative signaling dimer. J Biol Chem 283(18): 11861–11865
19. Baker D, Sali A (2001) Protein structure prediction and structural genomics. Science 294(5540):93–96
20. Arnold K, Bordoli L, Kopp J, Schwede T (2006) The SWISS-MODEL workspace: a web-based environment for protein structure homology modelling. Bioinformatics 22(2):195–201
21. Roy A, Kucukural A, Zhang Y (2010) I-TASSER: a unified platform for automated protein structure and function prediction. Nat Protoc 5(4):725–738
22. Wei T, Gong J, Rossle SC, Jamitzky F, Heckl WM, Stark RW (2011) A leucine-rich repeat assembly approach for homology modeling of the human TLR5-10 and mouse TLR11-13 ectodomains. J Mol Model 17(1):27–36
23. Wheeler DL, Barrett T, Benson DA, Bryant SH, Canese K, Chetvernin V, Church DM, Dicuccio M, Edgar R, Federhen S et al (2008) Database resources of the National Center for Biotechnology Information. Nucleic Acids Res 36(Database issue):D13–D21
24. Wei T, Gong J, Jamitzky F, Heckl WM, Stark RW, Rossle SC (2008) LRRML: a conformational database and an XML description of leucine-rich repeats (LRRs). BMC Struct Biol 8:47
25. Berman HM, Westbrook J, Feng Z, Gilliland G, Bhat TN, Weissig H, Shindyalov IN, Bourne PE (2000) The protein data bank. Nucleic Acids Res 28(1):235–242

26. Rice P, Longden I, Bleasby A (2000) EMBOSS: the European Molecular Biology Open Software Suite. Trends Genet 16(6):276–277
27. Fiser A, Do RK, Sali A (2000) Modeling of loops in protein structures. Protein Sci 9(9): 1753–1773
28. Fiser A, Sali A (2003) ModLoop: automated modeling of loops in protein structures. Bioinformatics 19(18):2500–2501
29. Wallner B, Elofsson A (2003) Can correct protein models be identified? Protein Sci 12(5): 1073–1086
30. McGuffin LJ, Roche DB (2010) Rapid model quality assessment for protein structure predictions using the comparison of multiple models without structural alignments. Bioinformatics 26(2):182–188
31. Laskowski RA, MacArthur MW, Moss DS, Thornton JM (1993) PROCHECK: a program to check the stereochemical quality of protein structures. J Appl Crystallogr 26:283–291
32. Humphrey W, Dalke A, Schulten K (1996) VMD: visual molecular dynamics. J Mol Graph 14(1):33–38, 27–38
33. Baker NA, Sept D, Joseph S, Holst MJ, McCammon JA (2001) Electrostatics of nanosystems: application to microtubules and the ribosome. Proc Natl Acad Sci U S A 98(18): 10037–10041
34. Maiti R, van Domselaar GH, Zhang H, Wishart DS (2004) SuperPose: a simple server for sophisticated structural superposition. Nucleic Acids Res 32(Web Server issue):W590–W594
35. Morris GM, Huey R, Lindstrom W, Sanner MF, Belew RK, Goodsell DS, Olson AJ (2009) AutoDock4 and AutoDockTools4: automated docking with selective receptor flexibility. J Comput Chem 30(16):2785–2791

Chapter 6

Nucleic Acid Recognition in Dendritic Cells

Alexander Heiseke, Katharina Eisenächer, and Anne Krug

Abstract

The immune system consists of specialized cell types with distinct functions in order to provide an effective innate and adaptive immune defense against harmful invading pathogens like bacteria, viruses, fungi, parasites, or other substances threatening the integrity of the organism. Once the immune system recognizes such pathogens via pattern recognition receptors (PRRs), they are taken up, processed, and presented as antigens on MHC class I and II to T lymphocytes by specialized cells called dendritic cells (DCs). At the same time pathogen components which bind to PRRs in DCs trigger potent cytokine and chemokine responses. Although other cell types like macrophages can also take up, process, and present antigens to naïve T lymphocytes, DCs are the cells with the greatest capacity to do so. Thus, DCs are also called professional antigen presenting cells (APCs), which induce a strong adaptive immune response and thereby act as a bridge between the innate and adaptive immune system.

This chapter provides detailed instructions on how to generate various types of DCs from human peripheral blood mononuclear cells (PBMCs) and murine bone marrow, as well as stimulation conditions for activation of these cells by PRR ligands in vitro.

Key words Dendritic cells, DCs, Plasmacytoid dendritic cells, pDCs, Monocyte-derived dendritic cells, Bone marrow-derived dendritic cells, Pattern recognition receptor, PRR, Toll-like receptor, TLR, Granulocyte macrophage colony-stimulating factor, GM-CSF, Interleukin-4, IL-4

1 Introduction

The immune system protects the host against invading harmful pathogens. It consists of two branches, the innate and the adaptive immune system. While the innate branch precedes the adaptive response and provides a fast defense mechanism against pathogens, the adaptive branch steps in later and generates long-lasting immunological memory. Thus, a faster adaptive immune response is achieved after the second encounter of a specific pathogen providing protective immunity.

Dendritic cells (DCs) play a special role in the interaction of the innate and the adaptive immune branches. DCs link both parts of the immune system by recognizing pathogens and triggering innate

Hans-Joachim Anders and Adriana Migliorini (eds.), *Innate DNA and RNA Recognition*, Methods in Molecular Biology, vol. 1169, DOI 10.1007/978-1-4939-0882-0_6,

immunity on the one hand, and by presenting antigens derived from these pathogens to T and B lymphocytes on the other hand.

The recognition of pathogens by DCs is carried out by specialized pattern recognition receptors (PRRs) [1, 2]. Four major groups of these PRRs have been identified: (a) toll-like receptors (TLRs), (b) retinoic acid inducible gene-1 like receptors (RLRs) [3], (c) nucleotide oligomerization domain (NOD)-like receptors (NLRs) [4], and (d) C-type lectins [5, 6]. Immediately after recognition of a molecular pattern by the designated PRR, DCs are activated to undergo a process called maturation. This encompasses the upregulation of costimulatory molecules and MHC class II as well as the production of cytokines and chemokines. In addition, DCs undergo morphological changes and start to migrate to the draining lymph nodes [7]. DCs constantly internalize, process, and present antigens on MHC class II to $CD4^+$ T cells [8, 9] and on MHC class I to $CD8^+$ T cells [9, 10]. Upon pathogen encounter and triggering of PRRs, the antigen processing and loading onto MHC molecules is greatly enhanced. In part this is due to increased and prolonged expression of peptide MHC complexes on the surface of DCs [11, 12].

Costimulatory molecules, for example molecules of the B7 family, such as CD80 and CD86 are upregulated which provide a required second signal for the naïve T cells to expand [13, 14]. Additionally, cytokine production by DCs is essential for subsequent differentiation of T cells into effector T helper and cytotoxic T cells [15–17].

Several functionally distinct DC subpopulations have evolved, providing optimized protection against invading pathogens. The two major subtypes are conventional DCs (cDCs) [18] and plasmacytoid DCs (pDCs) [19]. Both cDCs and pDCs develop from common progenitor cells in the bone marrow and require the growth factor FMS-like tyrosine kinase 3 ligand (Flt3-L) [20, 21]. Granulocyte-macrophage colony stimulating factor (GM-CSF) also drives cDC development but is not essential for cDC development in vivo [22]. Both, pDCs and cDCs, can be generated from murine bone marrow cells in vitro. For the development of cDCs, GM-CSF with or without addition of interleukin-4 (IL-4) is used [23, 24]. At present, it is unclear, if GM-CSF-DCs generated in this culture system truly have a counterpart in vivo. They most closely resemble monocyte-derived so called “inflammatory DCs” [25]. In bone marrow cell cultures with Flt3-L all subpopulations of pDCs and cDCs resembling the DC subset composition in the spleen can be generated [19, 26].

In the human setting, peripheral blood mononuclear cells (PBMCs) are used to generate monocyte-derived DCs. These DCs are generated like murine cDCs using GM-CSF and IL-4.

In the following sections, the materials and methods to generate these three types of DCs are provided, including specific comments on these protocols.

2 Materials

2.1 Human PBMC Isolation

2.1.1 Hardware and Media for PBMC and CD14-Positive Cell Isolation

1. PBS without Ca^{2+} and Mg^{2+}.
2. Red blood cell lysis buffer.
3. Ficoll.
4. Buffy coat.
5. MACS buffer: PBS without Ca^{2+} and Mg^{2+}, containing 2 % (v/v) FBS and 2 mM EDTA (store at 4 °C).
6. MACS pre-separation filters (30 μm).
7. MACS LS columns.
8. MACS stand.
9. CD14 Microbeads.

2.1.2 Dishes and Media for DC Generation

1. Complete DC medium: RPMI 1640 (*see* **Note 1**), supplemented with 10 % (v/v) FBS (*see* **Note 2**), 1 % (v/v) Penicillin [10,000 U/mL]/Streptomycin [10 mg/mL], 1 % (v/v) 100 mM Na-Pyruvate, 1 % (v/v) nonessential amino acids (NEAA) [100×], 1 % (v/v) 200 mM of L-alanyl-L-glutamine dipeptide, 50 μM β-mercaptoethanol.
2. rhGM-CSF.
3. rhIL-4.

2.2 Generation of Murine Bone Marrow-Derived DCs

2.2.1 Hardware

1. Scissors and forceps.
2. 10 cm petri dishes (better than tissue culture treated dishes).
3. Syringes (10 mL).
4. Disposable hypodermic needles (0.55 × 25 mm).

2.2.2 Media and Growth Factors

1. RPMI 1640.
2. Complete DC medium (*see* Subheading 2.1.2.1).
3. rmGM-CSF.
4. rmIL-4.
5. rhFlt-3 ligand.

3 Methods

3.1 PBMC Isolation

1. Prepare 15 mL Ficoll (room temperature, RT) per 50 mL Falcon tube.
2. Mix fresh blood 1:1 with *RT* PBS (*see* **Note 3**).
3. Layer approximately 20–25 mL of blood/PBS mixture onto the Ficoll (*see* **Note 4**).
4. Centrifuge at RT, 900 × *g*; no acceleration/no break for 20 min.

5. Carefully take out the PBMC containing white interphase using a 10 mL pipette and transfer into fresh 50 mL Falcon tube (*see* **Note 5**).
6. Centrifuge at 450×*g*, 4 °C for 10 min.
7. Wash pellet 3× (resuspend pellet in 10 mL PBS, add 35 mL PBS), centrifuge at 450×*g*, 4 °C for 5 min.
8. Resuspend pellet in 10 mL red blood cell lysis buffer, incubate at RT for 5 min (*see* **Note 6**).
9. Add complete DC medium to stop RBC lysis.
10. Centrifuge at 450×*g*, 4 °C for 5 min.
11. Resuspend pellet in complete DC medium.
12. Take an aliquot for counting cells after RBC lysis.
13. Wash cells once in PBS.
14. Resuspend cells in MACS buffer at the desired cell density.

3.2 Generation of Monocyte-Derived DCs from Human PBMCs

1. CD14 MACS (according to manufacturer's protocol; Miltenyi Biotech (*see* **Note** 7).
2. Wash cells after MACS isolation to remove EDTA and count cells.
3. Plate 3×10^6 $CD14^+$ cells per well (6-well plate) in 2 mL DC-Medium including 50 ng/mL GM-CSF and 1,000 U/mL IL-4 (*see* **Note 8**).
4. On day 4 of culture: add GM-CSF (25 ng/mL) directly to the wells; no additional IL-4 is needed.
5. On day 6–7: monocytes have differentiated into DCs.

See ***Note 9*** *for general comment on generation of monocyte-derived human DCs (Fig. 1).*

3.3 Isolation of Bone Marrow Cells from Mouse Hind Legs

1. Sacrifice mouse.
2. Take hind legs by cutting through the hip, so that the femur and tibia stay intact.
3. Remove all fur and muscles.
4. Under sterile conditions separate the femur from tibia by cutting through the knee joint.
5. Open femur and tibia at both ends, respectively.
6. Flush out bone marrow of femur and tibia with syringes containing RPMI 1640.
7. Prepare single cell suspension of the bone marrow by pipetting up and down several times with a 10 mL pipette.
8. Centrifuge single cell suspension at 450×*g*, 4 °C for 5 min.
9. Resuspend pellet in 1 mL red blood cell lysis buffer, incubate for 5 min at RT (*see* **Note 10**).

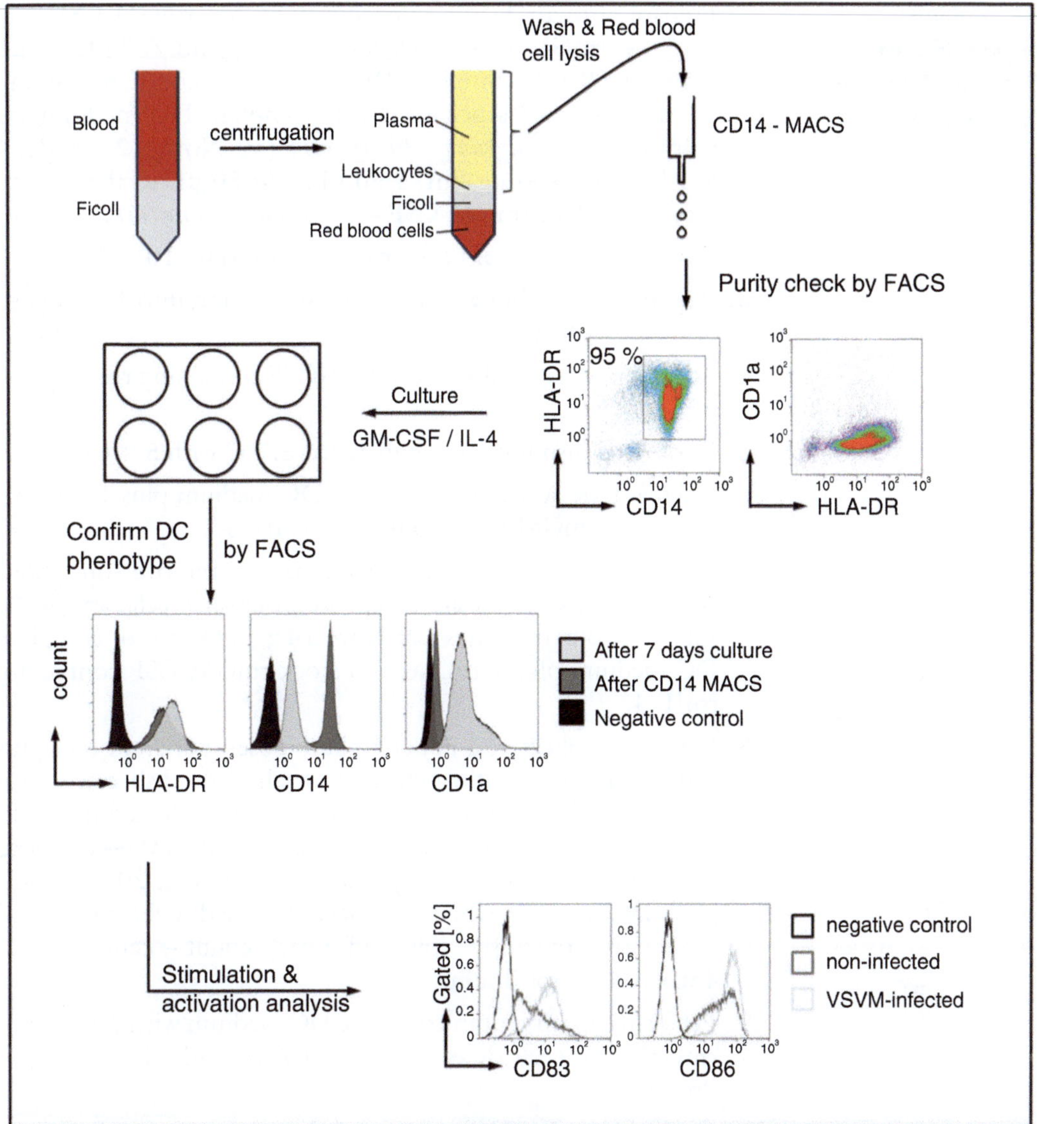

Fig. 1 Scheme for generating human monocyte-derived dendritic cells (DCs). After centrifugation of the blood, which was laid on top of the Ficoll, the plasma and leukocyte fractions are collected, washed, and lysed. Following monocyte isolation by MACS with anti-CD14 beads, the purity is checked by FACS. 6–7 days of monocyte culture with GM-CSF and IL-4 lead to the differentiation into DCs, which can be checked by FACS analysis, showing a downregulation of CD14 and an upregulation of the DC-specific marker CD1a. After stimulation (for example by VSV-M infection) the expected upregulation of the costimulatory molecules (like CD83 and CD86) can be measured by FACS analysis

10. Stop red blood cell lysis by adding 10 mL PBS containing 10 % (v/v) FBS (*see* **Note 11**).
11. Centrifuge single cell suspension at 450 × *g*, 4 °C for 5 min.
12. Resuspend pellet in 30–50 mL DC medium.
13. Count cells.

3.4 Culture of GM-CSF Expanded DCs from Murine Bone Marrow Cells

1. Prepare growth factor concentrations for rmGM-CSF and rm-IL-4: rmGM-CSF Stock conc.: 200 μg/mL/50 μL; 50 μL stock in 50 mL complete DC medium (second stock); add 1 mL of second stock to 9 mL complete DC medium for cultivation → *final conc.: 20 ng/mL;* (*see* **Note 12**); rmIL-4 Stock conc.: 200 μg/mL à 50 μL; add 50 μL to the 50 mL second stock of rmGM-CSF → *final conc.: 20 ng/mL.*
2. Adjust cell density of bone marrow cells to 6×10^5 cells/mL.
3. Pipette 9 mL of bone marrow cell suspension into 10 cm petri dish (*see* **Note 13**).
4. Add 1 mL of second stock rmGM-CSF containing rmIL-4 and mix gently.
5. Place petri dish into an incubator at 37 °C with 5 % CO_2.
6. After 3 days: add 9 mL complete DC medium plus 1 mL second stock rmGM-CSF, containing rmIL-4.
7. On day 5: discard 10 mL of supernatant (remove some non-adherent cells, if they are too dense); optional: cell can also be split in two petri dishes (*see* **Note 14**); add 9 mL of complete DC medium plus 1 mL second stock rmGM-CSF containing rmIL-4.
8. On day 7: collect non-adherent and loosely adherent cells plus supernatant by gentle pipetting; wash petri dish once with PBS, add this suspension to the previously collected one, add 5 mL of PBS-EDTA (final concentration: 500 μM) → incubate at 37 °C, about 5 min, stop reaction by adding 20 mL complete DC medium when cells have detached; collect cells and mix with previously collected ones, count → centrifuge at $450 \times g$, 4 °C for 5 min.
9. Perform experiment in complete DC medium with 20 ng/mL rmGM-CSF (cell density 0.5–1×10^6 per mL, 1–2×10^5 per 96 well).
10. *Optional for FACS analysis of DC activation markers* (*avoid "self-activation"!*): harvest cells on day 5 only by pipetting if possible (avoid centrifugation), and plate them in recommended concentration (1×10^6 cells/mL) in desired wells (24-well plate à 1 mL; or 96-well plate à 200 μL) in their medium plus some fresh complete DC medium with 20 ng/mL rmGM-CSF; culture them until day 7, directly add stimuli to wells and incubate for 24–36 h.
11. *Optional for collection of RNA from stimulated DCs*: add stimuli directly to 10 cm petri dish on day 7 of culture without disturbing the cells. After the desired time points (between 30 min and 12 h), collect the non-adherent cells in the supernatant in a 50 mL Falcon tube, collect remaining non-adherent cells with PBS, add them to a Falcon tube. Pipette TRIzol

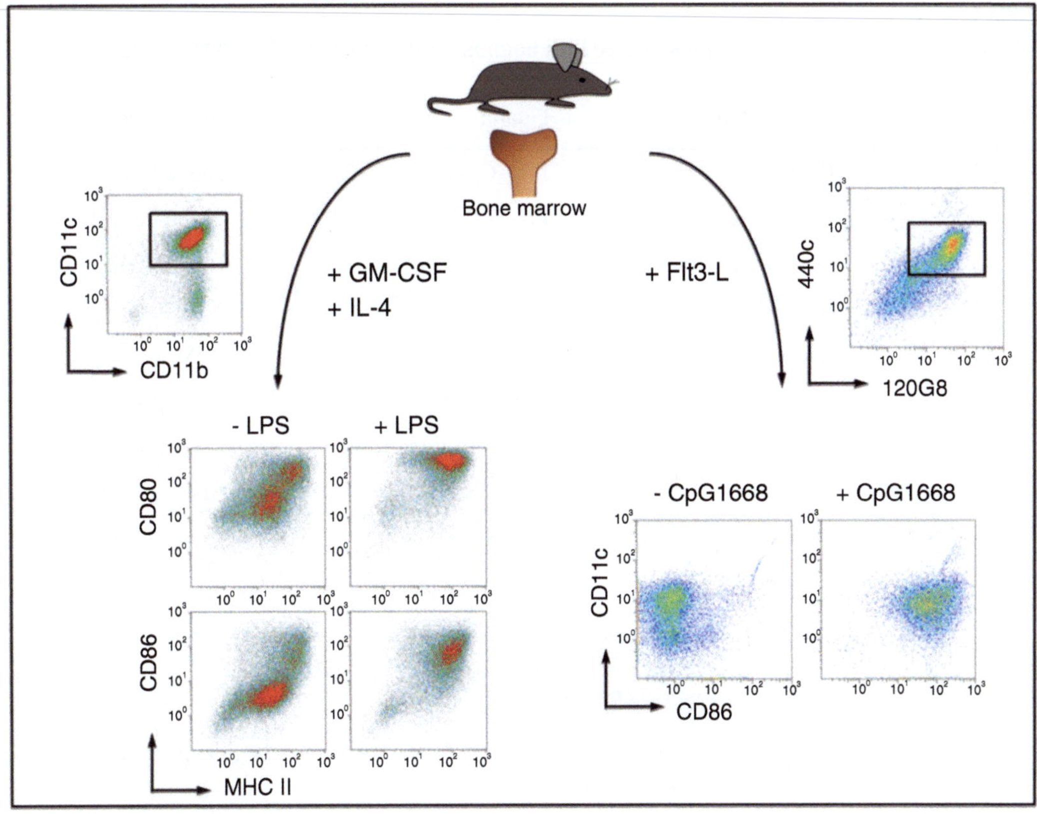

Fig. 2 Generation of cDCs and pDCs from murine bone marrow. After preparing single cell solutions of bone marrow cells, red blood cells are removed with red blood cell lysis buffer. The remaining cells are either treated with rmGM-CSF and rmIL-4 to generate cDCs, or with Flt3-L to generate cDCs and pDCs. GM-CSF-DC culture should result in at least 85 % pure CD11c/CD11b double positive cells, while Flt3L-DC culture should result in over 90 % CD11c^{+} cells containing BST2^{hi}/SiglecH^{hi} pDCs (40–60 %) and BST2^{lo}/SiglecH$^{neg/lo}$ cDCs. The pDCs express lower levels of MHC class II and costimulatory molecules than the cDCs, they are also CD11b^{lo}, and B220^{+}, while the cDCs are CD11b^{hi} and B220^{-} (not shown). After stimulation with TLR-ligands, cDCs and pDCs upregulate the expression of MHC class II, CD80 and CD86 surface molecules

Reagent (Invitrogen) directly onto adherent cells in the Petri dish and lyse cells by pipetting up and down. In the meantime spin down the cells in the Falcon tube, remove supernatant, keep on ice and then resuspend dry pellet in the TRIzol lysate. Transfer the lysate (from all cells) into a tube and store at −20 or −80 °C.

12. *Always*: check purity and activation status of DCs by FACS analysis before stimulation (*see* Fig. 2).

3.5 Culture of Flt-3 Ligand Expanded DCs from Murine Bone Marrow Cells

1. After isolation of bone marrow cells (*see* Subheading 3.3), take up cells in complete DC medium containing 20 ng/mL Flt3-L.
2. Adjust cell number to 1.5×10^{6} cells/mL.
3. Pipette 3 mL of bone marrow cell suspension into each well of a 6-well plate.

Table 1
Stimulation conditions of DCs with selected PRR ligands for cytokine ELISA as final readout

PRR	Localisation	Ligand	Ligand concentration
TLR2	Plasma membrane	Pam_3Cys	100–500 ng/mL
TLR3	Endosome	Poly(I:C)	20–50 μg/mL
TLR4	Plasma membrane	LPS	10–1,000 ng/mL
TLR5	Plasma membrane	Flagellin	0.1–10 μg/mL
TLR7/8	Endosome	R848	3 μM
TLR9	Endosome	CpG DNA	0.5–1 μM
RIG-I	Cytosol	5′ triphosphate RNA[a]	1–5 μg/mL
MDA-5	Cytosol	Poly(I:C)[a]	20–50 μg/mL

[a]Ligands need to be transfected (e.g., using Lipofectamine 2000, Invitrogen; according to the manufacturer's protocol)

4. Incubate cells for 7 days at 37 °C with 5 % CO_2.
5. Check purity and frequency of pDCs and cDCs by FACS analysis (*see* Fig. 2).

See ***Note 15*** *for general comments on the generation of murine bone marrow-derived DCs.*

3.6 Stimulation of DCs with TLR and RLR Ligands in Vitro

In general, the best condition for each stimulation and the respective application (e.g., cytokine ELISA, FACS, real-time qPCR, western blot) should be titrated and evaluated. In the following Table 1 of stimulation concentrations for selected PRR ligands are provided that have been evaluated in our lab to detect cytokines in the supernatant of stimulated DCs.

4 Notes

1. Depending on the manufacturer the yield of DCs and their purity can vary—if the purity is too low, or the numbers too few, various RPMI 1640 can be tested.
2. FBS can activate DCs in culture; therefore, heat-inactivating FBS can lower the activation status of DCs; however, heat inactivation of FBS might also lower the yield of DCs, and thus, it should be carefully considered. FBS differs from various manufacturers, and thus, several FBS should be tested if the yield or purity is not high enough.
3. Buffy coat, PBS and Ficoll should be used at RT, otherwise the cells will not separate.
4. Pipette very carefully, and do not mix blood with Ficoll.

5. Rather, take some parts of the plasma above the ring than contaminating the leukocytes with red blood cells—plasma will be discarded during washing steps.
6. Sometimes, red blood cell lysis has to be performed two times for complete lysis.
7. Always keep cells at 4 °C during incubation to avoid unspecific binding of Micro-Beads.
8. A lower cell density, e.g., 1×10^6 per well, is also possible and may reduce spontaneous maturation of DCs.
9. For PBMC isolation, Leucosep tubes (Greiner Bio One) can be used for Ficoll separation instead of Falcon tubes.

   It is crucial to avoid endotoxin contamination in the medium and in the serum. For some purposes (e.g., DC vaccine development), it may be beneficial to use medium supplemented with human serum or to use serum-free medium. These protocols have been published extensively elsewhere (refs. 27–29.)

   It was also shown that human monocyte-derived DCs can be generated with shorter incubation times ("fast DCs," *see* ref. 30.)

   Caution has to be used in extrapolating results obtained from monocyte-derived DCs to primary human DCs. Results should be confirmed using primary human cDC and pDC subpopulations isolated from PBMCs or other human tissues. In general, human monocyte-DCs constitute an excellent tool to perform functional studies, especially when techniques such as siRNA transfection are used.
10. 1 mL Red blood cell lysis buffer can be used for up to four legs (from two mice), for more cells the volume should be increased; sometimes, a second lysis might be necessary for complete lysis.
11. FBS is not necessary, but increases the survival of the cells.
12. rmGM-CSF is only stable for approximately 2 weeks at 4 °C and will rapidly lose activity once it is thawed, so prepare suitable aliquots in first freezing step to avoid multiple freezing–thawing of a single aliquot.
13. Most bone marrow-derived dendritic cells differentiated with rmGm-CSF are floating in the medium, thus, petri dishes with non-adherent surfaces should be used.
14. Avoid strong pipetting, as this might already activate the DCs.
15. To avoid activation of DCs, always prevent bubbles when pipetting; bubbles might also decrease the yield of DCs by killing cells.

    For both protocols it is important to test several batches of FBS to obtain optimal results with regard to DC yield, activation status and responsiveness to stimulation. Longer incubation times in the GM-CSF culture lead to a higher percentage of cDCs; however, these mature spontaneously in culture, so culture times should not be prolonged too much. Immature GM-CSF-DCs can be further purified by cell sorting if required (gating on $CD11c^{+}$ $CD11b^{+}$ $CD86^{lo}$ cells).

In the Flt3L-DC culture system, the percentage of pDCs is higher at earlier time points while the percentage of cDCs increases with longer incubation times. The cDC population contains at least two subpopulations which resemble those found in the spleen (*see* Naik et al. [26].). In the Flt3L-DC culture, the batch of the serum used has a great impact on the frequency of the pDC and cDCs subsets. In addition, it is important to maintain the cells in contact with the layer of adherent cells, which forms during the culture, since they act as natural feeder cells providing survival factors. Removal of the mostly non-adherent or loosely adherent Flt3L-DCs from the adherent cell layer and resuspension in fresh medium (with or without Flt3L) leads to restricted survival in further culture (maximum 36–48 h based on our experience).

In comparison to splenic DC subsets, we have found that pDCs generated from bone marrow cells with Flt3L culture in vitro are phenotypically and functionally similar to splenic pDCs but appear to be less differentiated, rather resembling primary pDCs which can be isolated from fresh murine bone marrow cells. Therefore, it may be necessary to confirm results by using primary pDCs isolated from spleen and/or lymph nodes.

References

1. Medzhitov R, Preston-Hurlburt P, Janeway CA Jr (1997) A human homologue of the Drosophila Toll protein signals activation of adaptive immunity. Nature 388:394–397
2. Poltorak A, He X, Smirnova I, Liu MY, Van Huffel C, Du X, Birdwell D, Alejos E, Silva M, Galanos C, Freudenberg M, Ricciardi-Castagnoli P, Layton B, Beutler B (1998) Defective LPS signaling in C3H/HeJ and C57BL/10ScCr mice: mutations in Tlr4 gene. Science 282:2085–2088
3. Yoneyama M, Kikuchi M, Natsukawa T, Shinobu N, Imaizumi T, Miyagishi M, Taira K, Akira S, Fujita T (2004) The RNA helicase RIG-I has an essential function in double-stranded RNA-induced innate antiviral responses. Nat Immunol 5:730–737
4. Inohara N, Ogura Y, Chen FF, Muto A, Nunez G (2001) Human Nod1 confers responsiveness to bacterial lipopolysaccharides. J Biol Chem 276:2551–2554
5. Yamasaki S, Matsumoto M, Takeuchi O, Matsuzawa T, Ishikawa E, Sakuma M, Tateno H, Uno J, Hirabayashi J, Mikami Y, Takeda K, Akira S, Saito T (2009) C-type lectin Mincle is an activating receptor for pathogenic fungus, Malassezia. Proc Natl Acad Sci U S A 106:1897–1902
6. Ogden CA, de Cathelineau A, Hoffmann PR, Bratton D, Ghebrehiwet B, Fadok VA, Henson PM (2001) C1q and mannose binding lectin engagement of cell surface calreticulin and CD91 initiates macropinocytosis and uptake of apoptotic cells. J Exp Med 194:781–795
7. Randolph GJ, Ochando J, Partida-Sanchez S (2008) Migration of dendritic cell subsets and their precursors. Annu Rev Immunol 26: 293–316
8. Villadangos JA (2001) Presentation of antigens by MHC class II molecules: getting the most out of them. Mol Immunol 38:329–346
9. Littman DR (1987) The structure of the CD4 and CD8 genes. Annu Rev Immunol 5: 561–584
10. Gromme M, Neefjes J (2002) Antigen degradation or presentation by MHC class I molecules via classical and non-classical pathways. Mol Immunol 39:181–202

11. Hiltbold EM, Safley SA, Ziegler HK (1996) The presentation of class I and class II epitopes of listeriolysin O is regulated by intracellular localization and by intercellular spread of Listeria monocytogenes. J Immunol 157: 1163–1175
12. Anderson HA, Hiltbold EM, Roche PA (2000) Concentration of MHC class II molecules in lipid rafts facilitates antigen presentation. Nat Immunol 1:156–162
13. Lenschow DJ, Walunas TL, Bluestone JA (1996) CD28/B7 system of T cell costimulation. Annu Rev Immunol 14:233–258
14. Herold KC, Vezys V, Koons A, Lenschow D, Thompson C, Bluestone JA (1997) CD28/B7 costimulation regulates autoimmune diabetes induced with multiple low doses of streptozotocin. J Immunol 158:984–991
15. Hsieh CS, Macatonia SE, Tripp CS, Wolf SF, O'Garra A, Murphy KM (1993) Development of TH1 CD4+ T cells through IL-12 produced by Listeria-induced macrophages. Science 260:547–549
16. Swain SL, Weinberg AD, English M, Huston G (1990) IL-4 directs the development of Th2-like helper effectors. J Immunol 145: 3796–3806
17. Veldhoen M, Hocking RJ, Atkins CJ, Locksley RM, Stockinger B (2006) TGFbeta in the context of an inflammatory cytokine milieu supports de novo differentiation of IL-17-producing T cells. Immunity 24:179–189
18. Henri S, Vremec D, Kamath A, Waithman J, Williams S, Benoist C, Burnham K, Saeland S, Handman E, Shortman K (2001) The dendritic cell populations of mouse lymph nodes. J Immunol 167:741–748
19. Gilliet M, Cao W, Liu YJ (2008) Plasmacytoid dendritic cells: sensing nucleic acids in viral infection and autoimmune diseases. Nat Rev Immunol 8:594–606
20. Waskow C, Liu K, Darrasse-Jeze G, Guermonprez P, Ginhoux F, Merad M, Shengelia T, Yao K, Nussenzweig M (2008) The receptor tyrosine kinase Flt3 is required for dendritic cell development in peripheral lymphoid tissues. Nat Immunol 9:676–683
21. Onai N, Obata-Onai A, Schmid MA, Ohteki T, Jarrossay D, Manz MG (2007) Identification of clonogenic common Flt3+M-CSFR+plasmacytoid and conventional dendritic cell progenitors in mouse bone marrow. Nat Immunol 8:1207–1216
22. Greter M, Helft J, Chow A, Hashimoto D, Mortha A, Agudo-Cantero J, Bogunovic M, Gautier EL, Miller J, Leboeuf M, Lu G, Aloman C, Brown BD, Pollard JW, Xiong H, Randolph GJ, Chipuk JE, Frenette PS, Merad M (2012) GM-CSF controls nonlymphoid tissue dendritic cell homeostasis but is dispensable for the differentiation of inflammatory dendritic cells. Immunity 36:1031–1046
23. Inaba K, Inaba M, Romani N, Aya H, Deguchi M, Ikehara S, Muramatsu S, Steinman RM (1992) Generation of large numbers of dendritic cells from mouse bone marrow cultures supplemented with granulocyte/macrophage colony-stimulating factor. J Exp Med 176: 1693–1702
24. Inaba K, Inaba M, Deguchi M, Hagi K, Yasumizu R, Ikehara S, Muramatsu S, Steinman RM (1993) Granulocytes, macrophages, and dendritic cells arise from a common major histocompatibility complex class II-negative progenitor in mouse bone marrow. Proc Natl Acad Sci U S A 90:3038–3042
25. Shortman K, Naik SH (2007) Steady-state and inflammatory dendritic-cell development. Nat Rev Immunol 7:19–30
26. Naik SH, O'Keeffe M, Proietto A, Shortman HH, Wu L (2010) CD8+, CD8-, and plasmacytoid dendritic cell generation in vitro using flt3 ligand. Methods Mol Biol 595:167–176
27. Thurner B, Roder C, Dieckmann D, Heuer M, Kruse M, Glaser A, Keikavoussi P, Kampgen E, Bender A, Schuler G (1999) Generation of large numbers of fully mature and stable dendritic cells from leukapheresis products for clinical application. J Immunol Methods 223: 1–15
28. Strasser EF, Eckstein R (2010) Optimization of leukocyte collection and monocyte isolation for dendritic cell culture. Transfus Med Rev 24:130–139
29. Schuler G, Steinman RM (1997) Dendritic cells as adjuvants for immune-mediated resistance to tumors. J Exp Med 186:1183–1187
30. Ramadan G (2012) In vitro expansion of human gammadelta and CD56(+) T-cells by Aspergillus-antigen loaded fast dendritic cells in the presence of exogenous interleukin-12. Immunopharmacol Immunotoxicol 34:309–316

Chapter 7

Viral Nucleic Acid Recognition in Human Nonimmune Cells: In Vitro Systems

Andrea Ribeiro and Markus Wörnle

Abstract

The innate immune system detects viral infections through a variety of receptors that sense the virus via their nucleic acid genome and evokes initial antiviral responses. Nonimmune cells possess pathogen-recognition mechanisms that enable them to respond to virus by the expression of RNA or DNA recognition receptors, thereby initiating the first steps in the host-pathogen interaction and inducing the production of pro-inflammatory and antiviral mediators. To better understand the activation of innate-immune defense mechanisms by nonimmune cells, we describe an experimental procedure that mimics viral infection, using four human nonimmune cell culture models under stimulation with synthetic analogues of RNA and DNA virus. Furthermore, we exemplify two viral infection models using cultured nonimmune cells and hepatitis C virus (HCV) or polyomavirus BK (BKV). In addition, we report the importance of siRNA technique in gene regulation in order to identify specific markers involved in innate antiviral responses in these cells.

Key words Nonimmune cells, PRRs, siRNA, Poly(I:C), Poly(dA:dT), HCV, BKV

1 Introduction

The innate immune system plays an essential role in recognizing viral infections via a variety of pattern recognition receptors (PRRs) that sense the virus through their nucleic acid genome or as a result of their replicative or transcriptional activity. The recognition of the virus by these receptors initiate a specific downstream signalling pathway leading to the activation of transcription factors, such as the nuclear factor kappa-light-chain-enhancer of activated B cells (NF-κB) or Interferon regulatory factors (IRFs) and induces pro-inflammatory and antiviral mediators including type-I IFNs and pro-inflammatory cytokines [1]. The production of these mediators is essential for an effective response against the virus.

Based on their cellular localization, PRRs involved in nucleic acid recognition can be divided in two groups: PRRs localized in the endosome and PRRs localized in the cytoplasm of the cell. The best-characterized classes of PRRs that participate in the recognition of

Hans-Joachim Anders and Adriana Migliorini (eds.), *Innate DNA and RNA Recognition*, Methods in Molecular Biology, vol. 1169, DOI 10.1007/978-1-4939-0882-0_7, © Springer Science+Business Media New York 2014

viral particles are Toll-like receptor (TLR) family and RIG-I-like receptor (RLR) family but lately it has become apparent that another series of PRRs called NOD-like receptors (NLRs) also play a role in viral innate immune response. Among TLRs, TLR3, TLR7, TLR8, and TLR9 have been shown to sense nucleic acids derived from virus in the endosome. Once TLRs are localized to endosome and are unable particles to recognize viral in the cytoplasm, RLRs family members such as RIG-I and MDA5 and NLRs family members as NLRP3, which are localized in the cytoplasm, function as alternative receptors for the recognition of viral particles in this site [2]. Recently it has also been described a member of the pyrin and HIN domain-containing protein (PYHIN) family, named AIM2, which senses foreign cytoplasmic DNA [3].

The cells involved in the innate immune response to the virus, which include dendritic cells (DCs) and macrophages, express a set of PRRs that are able to recognize conserved molecular patterns associated with the pathogen and induce a rapid inflammatory response stimulating the killing of infected or transformed cells, or stimulating phagocytosis or apoptosis [4]. However, specialized immune cells are often not the first cells to confront the virus. Nonimmune cells, such as epithelial cells or endothelial cells are also able to induce the production of antiviral and pro-inflammatory molecules by the expression of PRRs, playing a role in the first response against the pathogen as well as in the initiation of adaptive immune responses [5, 6].

Cell culture models have become a valuable tool used in cellular and molecular biology, providing excellent model systems for the study of cellular mechanisms under physiological and experimental conditions. Several authors documented a number of experiments performed in cell culture models in order to simulate viral infections and reproduce the mechanistic responses that may occur in vivo. Beside the description of experiments undertaken in specialized immune-cell models that are known to activate effective immune responses against the virus, many authors also reported efficient responses activated by nonimmune cells that can be potential targets for viral infections. For example, epithelial cells from the ocular surface are competent in activating an adequate immune response after being stimulated with a synthetic analogue of dsRNA virus, ***polyinosinic-polycytidylic acid (poly(I:C))*** [7]. Among other examples it was also demonstrated an involvement of lung epithelial cells in the immune response to dsRNA and influenza A virus [8].

Here we describe an experimental procedure which aims to simulate a viral infection in one model of primary cells (human primary mesothelial cells) and three cell-line models (human collecting duct cells, human mesangial cells (HMCs), and human microvascular endothelial cells (HMECs)), in order to study the subsequent innate immune response, by the stimulation of viral receptors and concomitant production of pro-inflammatory and antiviral mediators. Administration of the synthetic analogues of dsRNA (poly(I:C)) and dsDNA [polydeoxyadenylic-thymidylic acid (poly(dA:dT))] outside of the cells allows us to stimulate viral responses downstream to the activation of receptors located to endosome of the cell. Transfection technique was used to deliver the synthetic virus to the cytoplasm in order to activate the cytosolic receptors.

Furthermore, to simulate "in vivo" models of viral infection, we established two different in vitro infection models: cultured HMCs infected with HCV and human collecting duct cells infected with BKV.

In addition, we also present a protocol for knockdown experiments using short-interfering RNAS (siRNAs) that were conducted for the purpose of identifying which viral recognition receptors might contribute to the generation of antiviral and pro-inflammatory mediators.

2 Materials

2.1 Cell Culture

1. Buffer I (140 mM NaCl, 4 mM KCl, 11 mmol/L D-glucose, 1 % pyrogen-free human serum albumin, 10 mM HEPES (pH 7.3), 100 IU/mL penicillin, and 0.10 mg/mL streptomycin).
2. Growth medium I (MI99 medium supplemented with 10 % human serum, 10 % heat-inactivate new born calf serum, 25 mM Hepes (pH 7.3), 2 mM L-glutamine, 1 % penicillin–streptomycin).
3. Culture medium I (M199 medium supplemented with 2 % (vol/vol) human serum, 1 % penicillin–streptomycin).
4. Growth medium II (1:1 mixture of Dulbecco's Modified Eagle's medium and Ham's F12 medium supplemented with 5 μg/mL insulin, 5×10^{-8} M dexamethasone, 3×10^{-8} M sodium selenite, 5 μg/mL transferrin, 2 mM glutamine, 10 M HEPES buffer, 2 % heat-inactivated fetal bovine serum, and 1 % penicillin–streptomycin).
5. Growth medium III (Dulbecco's Modified Eagle's medium supplemented with 10 % heat-inactivated fetal bovine serum and 1 % penicillin–streptomycin).
6. Growth medium IV (Dulbecco's Modified Eagle's medium-low glucose containing 10 % of endothelial cell growth medium, 10 % of heat-inactivated fetal bovine serum, and 1 % penicillin–streptomycin).
7. Dulbecco's PBS (1×).
8. Trypsin–ethylenediaminetetraacetic acid (EDTA) (1×).
9. Fibronectin.

10. Collagenase type II.
11. Monoclonal antibodies against cytokeratin 8 and 18.

2.2 Stimulation Experiments

1. Poly(I:C).
2. Poly(dA:dT).
3. Lipofectamine 2000.
4. Opti-MEM I medium.

2.3 Knockdown Experiments with Short Interfering RNA (siRNA)

1. siPort NeoFX Transfection Agent.
2. Silencer predesigned siRNA for TLR3, RIG-I, and MDA5.
3. HP-Genome Wide siRNA for NLRP3 and AIM2.
4. Silencer negative control siRNA.

3 Methods

3.1 Isolation of Primary Human Mesothelial Cells (MCs)

1. MCs were isolated from omental tissue from patients undergoing elective surgery.
2. The tissue was collected in ice-cold buffer I and promptly transported to the laboratory.
3. The tissue was washed and cut in thin pieces into a tube I containing 0.1 % collagenase in M199 and incubated at 37 °C for 15 min.
4. After the incubation period, the pieces of tissue were transferred into a new tube (II) containing fresh M199 and gently washed.
5. The medium contained in both tube I and II was sieved through a 0.8-mm strainer into a new tube and centrifuged for 10 min at 1,000 RPM.
6. After centrifugation, the floating fat cells were carefully removed by aspiration, and so was the supernatant, and the pellet was resuspended in growth medium I.
7. Cells were grown in fibronectin-coated dishes in growth medium, at 37 °C under 5 % CO_2–95 % air atmosphere and the medium was replaced every 2–3 days.
8. When the cells had become confluent, subcultures were obtained by trypsin–EDTA treatment at a split ratio of 1:3.
9. For the experiments, the cells were used after 2 or 3 passages and the cells were always refed 48 h before the experiment with culture medium I.
10. To confirm the mesothelial nature, the cells were examined in their uniform cobblestone appearance at confluence and the abundant presence of cytokeratins 8 and 18 which are markers for mesothelial cells.

3.2 Cell Culture of Immortalized Human Collecting Duct Cells (HCDCs)

1. Immortalized HCDCs were cultured in growth medium II.
2. The cells were maintained under standard cultural conditions, at 37 °C supplied with 5 % CO_2.
3. When the cells became confluent, they were detached by treatment with a 1:1 mixture of trypsin–EDTA and PBS and were passage with a split ratio of 1:5.
4. Medium was renewed every 3 days.
5. 24 h before experiment, 2×10^5 cells were plated in a culture 6-well plate in 2 mL growth medium II so that cells will be around 90 % confluent at the time of stimulation.

3.3 Cell Culture of Immortalized Human Mesangial Cells (HMCs)

1. Immortalized HMCs were cultured in growth medium III and maintained in an incubator set at 37 °C supplied with 5 % CO_2.
2. When confluent, the cells were detached by treatment with trypsin–EDTA and passaged in a spilt ratio of 1:5.
3. The medium was replaced every 3 days.
4. One the day before experiment, 2×10^{-5} cells were plated in a culture 6-well plate with 2 mL of growth medium III.

3.4 Cell Culture of Immortalized Human Microvascular Endothelial Cells (HMECs)

1. Immortalized HMECs were maintained at 37 °C supplied with 5 % CO_2 in growth medium IV.
2. Subcultures were obtained by treatment with trypsin–EDTA at a split ratio of 1:5 and the medium was renewed every 3 days.
3. One the day before experiment, 3×10^{-5} cells were plated with 2 mL of growth medium IV in a culture 6-well plate.

3.5 Stimulation Experiments

3.5.1 Extracellular Poly(I:C) or Poly(dA:dT) Stimulation

1. After 24 h incubation, the medium was removed from the cells.
2. The cells were incubated with 2 mL of fresh culture medium alone (control) or medium containing poly(I:C) 10 μg/mL or poly(dA:dT) 10 μg/mL for 12 h at 37 °C under 5 % CO_2.
3. After incubation period, aliquots of the supernatant medium were removed for ELISA analysis and the cells were washed with PBS and prepared for RNA isolation.

3.5.2 Intracellular Poly(I:C) or Poly(dA:dT) Transfection

1. For the transfection, complexes of poly(IC) or poly(dA:dT) with the transfection agent (Lipofectamine 2000) were prepared in a ratio of 1 μg of RNA or DNA/1 μL of lipofectamine.
2. Poly(I:C) or poly(dA:dT) was diluted in 25 μL of Opti-MEM I reduced serum medium without serum and mixed gently.
3. The transfection agent was diluted in 25 μL of Opti-MEM I medium and incubated for 10 min at room temperature.
4. After incubation time the diluted RNA or DNA was combined with the diluted transfection agent, mixed gently, and incubated for 20 min at room temperature.

5. The medium from the cells prepared 24 h before was removed and 2 mL of fresh growth medium without antibiotics was added to the cells.
6. The complexes were added to each well containing cells and medium at a final concentration of 5 μg/mL and cells were incubated for 12 h at 37 °C under 5 % CO_2.
7. As a control, a complex of Opti-MEM medium mixed with the diluted transfection agent was used.
8. After 12 h, supernatants were removed for ELISA analysis and cells were prepared for RNA isolation.

3.5.3 Polyomavirus BK (BKV) Preparation and Infection of CDCs

1. BKV was isolated from a urine sample with a viral load of >10^8 copies/mL, from a patient with a known BKV reactivation.
2. The sample was inoculated for 1 h on Vero cells before minimal essential medium (MEM)/2 % FCS was added.
3. Viral propagation was monitored in the supernatant by quantitative real-time PCR with an in house-test on a 7500 Fast Real-Time PCR-System using primers from the Large T antigen region and a VIC-labelled probe.
4. At a load of 10^9 copies/mL, virus was harvested by centrifugation of the supernatant and subsequently passaged on HEK 293 cells with MEM/2 % FCS for 60 days.
5. Infected cells were split at 10-day intervals.
6. Viral growth was monitored by quantitative PCR and by immunofluorescence against T antigen.
7. To isolate BKV, HEK 293 cells were lysed by repeated freeze-thaw cycles (three times) and subsequent centrifugation at 800×g for 30 min.
8. The supernatant was adjusted to 5,000 FFU/mL with PBS and stored at −80 °C.
9. 24 h before infection, the cells were plated in a culture 6-well plate in 2 mL growth medium II.
10. For infection of the cells, virus was dissolved in growth medium II and 1 mL was added to HCDCs.
11. The cells were incubated for different time intervals (3, 6, 9, 12, and 24 h) at 37 °C supplied with 5 % CO_2. Growth medium alone was used as a control.
12. After each time point, corresponding supernatants were collected for ELISA analysis and the cells were prepared for RNA isolation.

3.5.4 Hepatitis C Virus (HCV) Preparation and Infection of HMCs

1. For HCV infection of the cells, HMCs were plated in culture six-well plates 1 day before the experiment.
2. Exosomes containing viral RNA were isolated from human sera of patients with high viral loads by centrifugation.

3. 20–50 mL of clotted human whole blood was centrifuged at 1,500 × *g* for 10 min.
4. The supernatant was collected and centrifuged again at 10,000 × *g* for 30 min to remove solid remnants.
5. The resulting supernatant was centrifuged at 70,000 × *g* for 60 min in a SW28 rotor.
6. The pellet was dissolved in 1 mL cell culture medium without serum or antibiotics and combined with Lipofectamine 2000.
7. Medium alone (control), or medium containing the virus was added to the cells, and the plates were centrifuged at 1,000 × *g* for 45 min to allow for an efficient virus infection.
8. Cells were incubated at 37 °C under 5 % CO_2 for different time intervals (3, 6, 12, and 18 h) and after each time interval supernatants were collected for ELISA analysis and cells were prepared for RNA isolation.
9. Concentration of HCV used for stimulation was 100×10^6 geq/mL confirmed by RT-PCR.

3.6 Knockdown Experiments with Short Interfering RNA (siRNA)

1. Knockdown experiments were conducted using siPort NeoFX transfection agent in combination with predesigned siRNAs for TLR3, RIG-I, MDA5, NLRP3, and AIM2 or in combination with a scrambled siRNA that was used as a nonspecific negative control.
2. Cells were trypsinized before transfection using the routine procedure.
3. Trypsin was inactivated by resuspending the cells with growth medium and cells were set aside at 37 °C while the transfection complexes were being prepared.
4. 5 μL of transfection agent was diluted in 95 μL of Opti-MEM I medium and incubated at room temperature for 10 min.
5. The small RNA was diluted in Opti-MEM I medium for a final concentration of 10 nM (that is the final concentration after the transfection complexes are mixed with the cells in **step 8**).
6. Diluted transfection agent was combined with the diluted RNA, mixed by pipetting up and down, and incubated at room temperature for 10 min.
7. The RNA–transfection agent complexes were dispensed into the empty wells of a culture 6-well plate.
8. The cells prepared previously were seeded into the culture plate wells containing the transfection complexes at a density of 2.3×10^5 cells/well.
9. The plate was gently rocked back and forth to evenly distribute the complexes and the cells were incubated at 37 °C in normal cell culture conditions for 24 h.

10. The following day, the medium containing the complexes was removed and new growth medium without (control) or with poly(I:C) or poly(dA:dT) was added to cells for 12 h as described above.
11. The supernatants were collected for ELISA analysis, and the cells were prepared for RNA isolation.

4 Notes

1. *To ensure the quality of the cells in culture*
   (a) The cells were exposed to trypsin–EDTA long enough to detach the cells, as a long exposition can damage cell surface receptors.
   (b) The cells were handled gently and centrifuged at low speed (1,000 RPM).
   (c) The cells were always subcultured at around 80–90 % confluence.
   (d) The cells were tested periodically for the presence of mycoplasma, and the cultures and medium were examined regularly for evidence of bacterial or fungal contamination.
   (e) All the cell culture experiments were performed in aseptic conditions.
2. *To ensure the quality of the stimulation experiments*
   (a) Poly(I:C) or poly(dA:dT) were aliquoted after resuspension to avoid repeated freeze–thaw cycles that can decrease the efficiency of the product.
   (b) Antibiotics-free medium was used during transfection as this can cause cell death.
   (c) To obtain the highest transfection efficiency and low cytotoxicity, optimization experiments were performed by varying the concentration of the transfection agent or the RNA or DNA.
3. *To ensure the quality of the knockdown experiments*
   (a) siRNAs were aliquoted in suitable amounts after resuspension to avoid repeated freeze-thaw cycles that can reduce the efficiency of the product.
   (b) Optimization experiments were performed to establish a balance between the highest knockdown efficiency and the lowest cytotoxicity. Different volumes of the transfection agent were tested, and so were different concentrations of siRNA. Different times of exposition to the complexes were also evaluated.

References

1. Takeushi O, Akira S (2009) Innate immunity to virus infection. Immunol Rev 227:75–86
2. Kawai T, Akira S (2009) The roles of TLRs, RLRs and NLRs in pathogen recognition. Int Immunol 21:317–337
3. Hornung V, Ablasser A, Charrel-Dennis M, Bauernfeind F, Horvath G, Caffrey DR, Latz E, Fitzgerald KA (2009) AIM2 recognizes cytosolic dsDNA and forms a caspase-1 activating inflammasome with ASC. Nature 458:514–518
4. Akira S, Uematsu S, Takeuchi O (2006) Pathogen recognition and innate immunity. Cell 124:783–801
5. Abreu MT, Fukata M, Arditi M (2005) TLR signaling in the gut in health and disease. J Immunol 174:4453–4460
6. Fritz JH, Le Bourhis L, Magalhaes JG, Philpott DJ (2008) Innate immune recognition at the epithelial barrier drives adaptive immunity: APCs take the back seat. Trends Immunol 29: 41–49
7. Ueta M, Hamuro J, Kiyono H, Kinoshita S (2005) Triggering of TLR3 by polyI:C in human corneal epithelial cells to induce inflammatory cytokines. Biochem Biophys Res Commun 331: 285–294
8. Guillot L, Le Goffic R, Bloch S, Escriou N, Akira S, Chignard M, Si-Tahar M (2005) Involvement of toll-like receptor 3 in the immune response of lung epithelial cells to double-stranded RNA and influenza A virus. J Biol Chem 280:5571–5580

Chapter 8

Analysis of Nucleic Acid-Induced Nonimmune Cell Death In Vitro

Simone Romoli and Adriana Migliorini

Abstract

Foreign nucleic acids are recognized by germ-line-encoded receptors expressed in immune and nonimmune cells. Activation of the nucleic acid-specific pattern recognition receptors by foreign nucleic acid promotes production of inflammatory cytokines (mostly type I IFNs) and at the later stage leads to cell death. Here, we describe reliable and simple methods to quantify cell death caused by nucleic acid recognition. Additionally, we report two different methods to discriminate between two cell death modalities: apoptosis and necrosis.

Key words Cell death, Apoptosis, Necrosis, Annexin, Flow cytometry

1 Introduction

Host- and pathogen-associated nucleic acids are sensed and recognized by innate immune-recognition receptors, known as PRRs (pattern recognition receptors), which activate antiviral pathways. Innate immune-recognition receptors (PRRs) are germ-line-encoded receptors expressed on the cell surface or in the intracellular compartment, in immune and nonimmune cells [1, 2]. In contrast to immune cells, somatic cells express a limited repertoire of extracellular and intracellular PRRs. The nucleic acid in viral genomes can be either DNA or RNA, single-stranded or double-stranded, positive or negative in polarity, and are mostly recognized by intracellular PRRs, localized in endosomal compartments. Cytosolic ssRNA is exclusively recognized by TLR 7/8, whereas dsRNA is recognized by TLR3 or by RIG I helicases such as RIG-I and MDA5 [3, 5]. TLR9 recognizes and responds to foreign nonmethylated CpG-containing island, dsDNA, by contrast, is recognized by AIM2 inflammasome [3, 4]. Activation of the viral pattern recognition receptors (antiviral pathways) by foreign nucleic acid promotes production of inflammatory cytokines (mostly type I IFNs) and at the later stage leads to cell death [2, 5].

Hans-Joachim Anders and Adriana Migliorini (eds.), *Innate DNA and RNA Recognition*, Methods in Molecular Biology, vol. 1169, DOI 10.1007/978-1-4939-0882-0_8, © Springer Science+Business Media New York 2014

Recently several forms of cell death have been defined, based on functional and morphological criteria. According to the recent classification, 13 cell death modalities (typical and atypical) have been reported [6]. However, despite the fact that cells can die in several ways, viral infections generally induce the classical and well-known cell death modalities: apoptosis and necrosis [7, 8]. Here we describe a reliable and simple method to estimate cell viability by nucleic acid stimulation in a primary cell model (murine mesangial cells) and in a non replicating cell line model, known as immortalized murine podocytes (MPC) [9]. Determining cytotoxic effects triggered by nucleic acid is a start point to evaluate activation of cell death modality pathways, such as apoptosis or necrosis.

Apoptosis is a homeostatic programmed cell death modality occurring in various physiological and pathological situations [10]. Apoptosis is a multi-pathway cell-death program characterized by morphological and biochemical hallmarks, including nuclear DNA fragmentation, cell shrinkage, and cell membrane changes [11]. Here, we describe an apoptosis assay based on the estimation of characteristic cell membrane changes of this programmed cell death. Viable cells are characterized by an asymmetric distribution of different phospholipids between the inner and outer cell membrane surface [12]. During early apoptosis, cell membrane changes are accompanied by exposure of phosphatidylserine, which is normally located in the inner membrane surface, to the outer membrane surface, and becomes available to be detected by Annexin V protein [13, 14]. At this stage DNA and chromatin structures are unchanged. Annexin V is member of Annexin proteins that specifically bind to phosphatidylserine in the presence of calcium [15]. In the later stage, DNA cleavage and chromatin condensation take place, leading to formation of apoptotic cell bodies, which are rapidly phagocytized and digested by immune cells [16]. Using propidium iodide (PI), an intercalating agent binding to nucleic acids, in addition to Annexin V, enables to distinguish between early apoptotic cells (Annexin V+/PI−), late apoptotic (Annexin V+/PI+), and necrotic cells (Annexin V−/PI+). In contrast to apoptosis, necrosis is a nonphysiological cell death pathway characterized by cell lyses without formation of vesicles. As a consequence, cellular contents are delivered into the intracellular space, causing damage to neighboring cells [17]. Here, we describe a step-by-step protocol based on the leakages of the acytoplasmic enzyme, lactate dehydrogenase (LDH), from necrotic cells [18]. To detect LDH release into culture supernatants, a microtiter plate-based colorimetric absorbance assay is described, in order to evaluate necrosis in response to nucleic acid stimulation.

2 Materials

2.1 Cell Culture and Stimulation Experiments

1. Dulbecco's 1× PBS.
2. Petri dishes.
3. Steel cell sieves: 45–50–63–106–150 μm.
4. 1, 2, 5, 50 mL syringe.
5. RPMI 1640 medium.
6. 70 μm plastic cell strainer.
7. 30 μm plastic cell strainer.
8. Surgical sterile scalpels and scissors.
9. Collagenase IV.
10. 1× Trypsine/Ethylenediaminetetraacetic acid (TE).
11. Penicillin–Streptomycin (PS).
12. Heat-inactivated Fetal Bovine Serum (FBS).
13. Insulin–Transferrin–Sodium Selenite Supplement (ITS).
14. Murine interferon gamma (mIFN-γ) (*see* **Note 1**).
15. Complete growth medium (RPMI 1640 medium supplemented with 20 % FBS, 1 % PS and 1 % ITS).
16. Growth medium (RPMI 1640 medium supplemented with 10 % FBS, 1 % PS).
17. Permissive growth medium (RPMI 1640 supplemented with 10 % FCS, 1 % PS and 100 U/mL mIFN-γ).
18. Non-permissive growth medium (RPMI 1640 supplemented with 10 % FCS, 1 % PS).
19. Collagen type I.
20. Poly (I:C).
21. Poly (dA:dT).
22. CpG ODNs.
23. Lipofectamine 2000.

2.2 Cells Detachment Assay

1. Trypan blue solution 0.4 %.

2.3 Apoptosis Assay

1. Fluorescent labeled Annexin V.
2. Propidium Iodide (PI).
3. Annexin V/PI buffer (0.1 M Hepes/NaOH (pH 7.4), 1.4 M NaCl, 25 mM $CaCl_2$).

2.4 Necrosis Assay

1. LDH assay (Kit from company).
2. 96-well plates.
3. Triton X-100 solution.

3 Methods

Carry out all the procedures under sterile condition unless otherwise specified.

3.1 Isolation of Primary Murine Mesangial Cells from Kidney

1. Collect kidneys from sacrificed animals and keep them in ice-cold PBS.
2. Remove the renal capsule and the medulla (the innermost part) from the kidneys using surgical instruments.
3. Transfer the kidneys in a petri dish containing 5 mL of PBS and cut the organ into thin pieces to form a "kidney paste" (*see* **Note 2**).
4. Build the "cell steel tower" in the following order (from the bottom to the top): 45–50–63–106–150 μm (*see* **Note 3**).
5. Sieve the kidney paste through 150 μm cell steel sieve using the inner part of a 2 mL syringe as pestle.
6. Wash the 150 μm cell steel sieve five times with 50 mL of PBS using a 50 mL syringes to apply pressure (*see* **Note 4**).
7. Remove (gently) the sieve and wash the bottom surface with 50 mL of PBS to collect the remaining glomeruli on the 106 μm cell steel sieve.
8. Repeat **steps 5–7**.
9. Remove the 106 μm sieve and wash one time the 63 μm steel sieve with 50 mL of PBS using a 50 mL syringe to apply pressure.
10. Remove the 63 μm steel sieve and wash the surface and the bottom of the sieve with 50 mL of RPMI 1640 medium supplemented with 10 % FBS to collect the remaining glomeruli in a falcon.
11. Repeat the **steps 9–10** for the 50 μm steel sieve and as well for the 45 μm steel sieve.
12. Centrifuge the collected glomerular suspension for 7 min at $400 \times g$ at 4 °C.
13. After centrifugation, remove the supernatant by aspiration and resuspend the pellet in 3 mL of RPMI 1640 medium supplemented with 10 % FBS.
14. Sieve the kidney suspension through a 70 μm cell strainer using the inner part of a 2 mL syringe as pestle and collect the suspension in a new falcon.
15. Repeat **step 13** using a 30 μm cell strain and using the inner part of a 1 mL syringe as pestle.
16. Overturn the 30 μm cell strain and rinsed with 10 mL of collagenase solution to collect the stuck glomeruli.

17. Dilute the collected glomeruli in RPMI 1640 medium supplemented with 1 mg/mL collagenase IV and incubate for 20 min at 37 °C.
18. Dilute the digested glomeruli with 20 mL of RPMI 1640 medium supplemented with 10 % FBS and centrifuged as mention in **step 11**.
19. After centrifugation, carefully remove the supernatant by aspiration and resuspend the pellet in pre-warmed complete growth medium (*see* **Note 5**).
20. Distribute and grow the isolated kidney cells in 6-well dishes in complete growth medium, at 37 °C under 5 % CO_2/95 % air atmosphere.
21. Add fresh complete growth medium every 2–3 days.
22. After 5 days, remove the old medium by aspiration and replace it with fresh medium.
23. Repeat **step 22** every 3 days.
24. After 16–20 days subculture are obtained by T/E treatment at a split ratio of 1:2.
25. After 5–6 passages replace the complete growth medium with growth medium.
26. After passage 10 stain the cells for mesangial markers (like α-sma or fibronectin) to confirm their mesangial nature and use the positive populations for experiments.

3.2 Cell Culture of Immortalized Murine Podocytes

1. Seed Immortalized murine podocytes MPCs in 10 cm plates coated with type I collagen and culture these cells in permissive growth medium at 33 °C supplied with 5 % CO_2 (*see* **Note 6**).
2. When the cells reach 70 % confluence, detach the cells by treatment with a 1:1 mixture of T/E and PBS.
3. To induce their differentiation, seed 7×10^4 cells per 10 cm plates coated with type I collagen and grown in non-permissive growth medium at 37 °C (restrictive grown condition), for 14 days (*see* **Note 7**).
4. After 2 weeks of culture under restrictive growth condition, fully differentiated mouse podocyte display an arborized shape expressing synaptopodin (podocyte marker). Cells exhibiting these features are suitable for experiments.

3.3 Cells Detachment Assay

1. Prepare fully differentiated MPC in 6-well plate at the confluence of 70 %.
2. For extracellular stimulation by nucleic acids, incubate the cells with medium alone (control) or medium containing poly (I:C), poly (dA:dT), or ODN CpG in increasing concentration for different time points.

3. For intracellular stimulation by nucleic acids, incubate the cells with medium containing increasing concentration of poly (I:C), poly (dA:dT), and CpG ODN complexes with the transfection agent (Lipofectamine 2000) at 1:1 (v/v) ratio or medium containing equivalent increasing concentration of transfection reagent (control) for different time points.
4. At the end of the first time point, collect supernatant from each stimulations and controls and centrifuged at 400×*g* for 10 min.
5. Simultaneously collect the remaining adherent cells by treatment with a 1:1 mixture of TE and PBS and centrifuge at 400×*g* for 10 min.
6. Resuspend respectively both the collected cellular pellet in 0.5 mL of PBS.
7. Mix 10 μL of each cellular suspension with 10 μL of trypan blue.
8. By the use of Neubauer's chamber, count the total number of trypan blue positive and negative cells.
9. Determinate the percentage of detached cells considering trypan blue positive cells counted in the supernatant over the total number of trypan blue negative cells (attached cell + detached cells) (*see* **Note 8**).

3.4 Apoptosis Assay

1. Seed cells in 6-well plate at the confluence of 70 %.
2. Stimulate the cells as described in **steps 2** and **3** of Subheading 3.3.
3. At the end of the first time point, collect supernatant from each stimulations and controls (*see* **Note 9**).
4. Simultaneously collect and the remaining adherent cells by treatment with a 1:1 mixture of TE (*see* **Note 10**).
5. Mix Cell suspension from each stimulation and control, together with the equivalent collected supernatant and centrifuge at 400×*g* for 10 min.
6. After centrifugation, remove carefully the supernatant by aspiration, and resuspend the pellet in 5 mL of Annexin V/PI buffer.
7. Centrifuge at 400×*g* for 10 min.
8. After centrifugation, resuspend the pellet in 100 μL of mix containing 5 μL of fluorescent labelled Annexin V, 5 μL PI and 90 μL of Annexin V/PI buffer.
9. Incubate the cells for 15–20 min in dark at room temperature (*see* **Notes 11** and **12**).
10. At the end of the incubation, dilute the cells in 400 μL of Annexin V/PI buffer and analyze by flow cytometry.

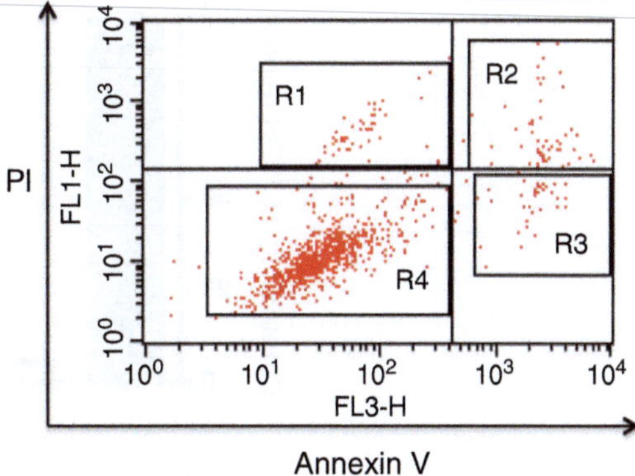

Fig. 1 Example of a FACS dot plot for apoptosis assay. (R1) Annexin V–/PI+ are necrotic cells, (R2) Annexin V+/PI+ are late apoptotic cells, (R3) Annexin V+/PI– are early apoptotic cells, and (R4) Annexin V–/PI– are viable cells

11. Determinate the percentage of apoptotic cell was determinate by comparing the percentage of the total cells Annexin V+/PI– founded in the stimulation to the equivalent control (Fig. 1).

3.5 Necrosis Assay

1. Seed cells in 96-well plate at the confluence of 70 %.
2. Stimulate the cells in triplicate as described in **steps 2** and **3** of Subheading 3.3.
3. Prepare Additional controls (according with the kit instruction) for both the intracellular and extracellular stimulation by nucleic acid: background control (assay medium without cells), spontaneous LDH release (assay medium containing cells) (low control), maximum LDH release (2 % tritonX-100 in assay medium containing cells) (high control).
4. At the end of the first time point, collect and transfer 100 μL supernatant from each stimulations and controls in a new 96-well plates (*see* **Note 13**).
5. Carefully add 100 μL of reaction mixture (freshly prepared) to the collected supernatants and incubate in the dark for 30 min at room temperature.
6. At the end of the incubation, measure the absorbance of the samples at 490–492 nm using a spectrophotometer (Fig. 2).
7. According with the kit instructions, calculate the average absorbance of the triplicate and subtract from each of these the absorbance value obtain in the background control and use the following formula to calculate the LDH activity (% cytotoxicity).

$$\text{Cytotoxicity}(\%): \frac{(\text{exp. value} - \text{low control})}{(\text{high control} - \text{low control})} \times 100.$$

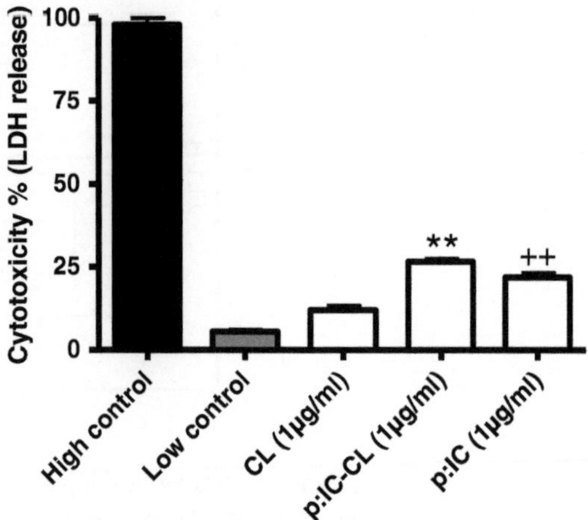

Fig. 2 Example of necrosis induced by nucleic acid. Cells were stimulated with nucleic acid for 24 h and LDH activity (% cytotoxicity) was measured. CL: Lipofectamine, p:IC-CL: poly (I:C) complexes Lipofectamine, p:IC: poly(I:C). ***P*< 0.01 *versus CL*; ++*P*< 0.001 *versus low control*

4 Notes

1. De-freeze a new IFN-gamma aliquot from −20 °C after the old aliquot at 4 °C is finished. Keep the de-frozen aliquot at 4 °C for 2–3 weeks only.
2. For makes "kidney paste" uses two parallel sterilized sharp scalpels.
3. It is a good practice to autoclave the sieves 1 day before the experiment.
4. It is recommended to apply high pressure on the sieves using 50 mL syringe to remove the tubular segments from the glomerular part.
5. During this step it is really important to observe and quantify the amount of glomeruli pellet in the bottom part of the falcon. If it is very small, it is better to add less quantity of medium; around 500 μL for each well. If the pellet is big enough; add 1 mL medium for each well. At the end, add 1 mL of fresh warm medium to make the final volume around 2 mL.
6. Prepare collagen I coated dishes at least 1 day before the experiment. Use 3–4 mL of collagen I solution (sterilized) for every 10 cm dish. After pouring all coating solution on the plate surface, keep plates at 37 °C for 10–15 min. After solidification of collagen I, remove excess (non-solid) collagen I solution with vacuum pump. Dry plates under cell culture hood by keeping them open for around 1 h. Switch off the light during all procedures, since collagen I is light sensitive.

7. Change the old non-permissive medium with 10 mL fresh medium after 1 week. This will increase the cell viability.
8. Do all these procedures as fast as possible to reduce the detachment of cells. It is recommended to count detached cells in a double blind manner.
9. After apoptosis stimulus, we can consider the cells that are in the supernatant as "late apoptotic cells." They have lost the capacity to stay attached on the plate.
10. Detach the remaining cells using TE. We can consider these cells as "early apoptotic cells." This way it is possible to do a complete FACS apoptosis assay.
11. The Annexin V/PI incubation at room temperature has a higher cell death index. On the other hand the capacity of Annexin V/PI entering in the cells is more sufficient compared to 4 °C incubation.
12. During this incubation step it is possible to do a co-incubation of Annexin V/PI with another antibody against a third target for screening cell subpopulations.
13. For LDH assay, it is always better to use fresh samples. Preserved samples (either at –20 or 4 °C) significantly alter the assay results.

References

1. Medzhitov R, Janeway CA Jr (1997) Innate immunity: impact on the adaptive immune response. Curr Opin Immunol 9(1):4–9
2. Janeway CA Jr, Medzhitov R (2002) Innate immune recognition. Annu Rev Immunol 20:197–216
3. Fernandes-Alnemri T et al (2009) AIM2 activates the inflammasome and cell death in response to cytoplasmic DNA. Nature 458(7237):509–513
4. Kumagai Y, Takeuchi O, Akira S (2008) TLR9 as a key receptor for the recognition of DNA. Adv Drug Deliv Rev 60(7):795–804
5. Kawai T, Akira S (2009) The roles of TLRs, RLRs and NLRs in pathogen recognition. Int Immunol 21(4):317–337
6. Galluzzi L et al (2012) Molecular definitions of cell death subroutines: recommendations of the Nomenclature Committee on Cell Death 2012. Cell Death Differ 19(1):107–120
7. Cho YS et al (2009) Phosphorylation-driven assembly of the RIP1-RIP3 complex regulates programmed necrosis and virus-induced inflammation. Cell 137(6):1112–1123
8. Tran AT et al (2013) Influenza virus induces apoptosis via BAD-mediated mitochondrial dysregulation. J Virol 87(2):1049–1060
9. Durvasula RV, Petermann AT, Hiromura K, Blonski M, Pippin J, Mundel P, Pichler R, Griffin S, Couser WG, Shankland SJ (2004) Activation of a local tissue angiotensin system in podocytes by mechanical strain. Kidney Int 65(1):30–39
10. Hengartner MO (2000) The biochemistry of apoptosis. Nature 407(6805):770–776
11. Igney FH, Krammer PH (2002) Death and anti-death: tumour resistance to apoptosis. Nat Rev Cancer 2(4):277–288
12. Bretscher MS (1972) Asymmetrical lipid bilayer structure for biological membranes. Nat New Biol 236(61):11–12
13. Koopman G et al (1994) Annexin V for flow cytometric detection of phosphatidylserine expression on B cells undergoing apoptosis. Blood 84(5):1415–1420
14. Martin SJ et al (1995) Early redistribution of plasma membrane phosphatidylserine is a general feature of apoptosis regardless of the

initiating stimulus: inhibition by overexpression of Bcl-2 and Abl. J Exp Med 182(5): 1545–1556
15. Tait JF, Gibson D, Fujikawa K (1989) Phospholipid binding properties of human placental anticoagulant protein-I, a member of the lipocortin family. J Biol Chem 264(14): 7944–7949
16. Kerr JF, Wyllie AH, Currie AR (1972) Apoptosis: a basic biological phenomenon with wide-ranging implications in tissue kinetics. Br J Cancer 26(4):239–257
17. Leist M, Jaattela M (2001) Four deaths and a funeral: from caspases to alternative mechanisms. Nat Rev Mol Cell Biol 2(8): 589–598
18. Chan FK, Moriwaki K, De Rosa MJ (2013) Detection of necrosis by release of lactate dehydrogenase activity. Methods Mol Biol 979:65–70

Chapter 9

In Vitro Analysis of Nucleic Acid Recognition in B Lymphocytes

Saskia Ziegler and Isabelle Bekeredjian-Ding

Abstract

In contrast to murine B cells, Toll-like receptor (TLR) expression in human B cells is mainly restricted to endosomally localized TLR7 and -9, receptors for RNA and DNA, respectively. Most importantly, B lymphocytes lack classical phagocytic receptors and instead internalize antigen only via the B cell receptor (BCR), a surface immunoglobulin specific for a defined antigen. BCR ligation triggers internalization of particulate antigens and physically associated molecules among them bacterial DNA or RNA. Thereby, this process provides access to endosomal nucleic acid-sensing TLRs. Co-stimulation of BCR and TLR ultimately leads to T cell-independent B cell activation. Here, we explain how this process can be experimentally mimicked in human peripheral blood B cells, e.g., using a microsphere-based system that promotes uptake of nucleic acid-based TLR ligands via BCR engagement.

Key words B cells, TLR, BCR, Mitogens, Proliferation, Differentiation

1 Introduction

B lymphocytes represent an essential component of adaptive immunity. They are specialized phagocytes that exclusively ingest antigen that is recognized by the B cell receptor (BCR), a surface-bound immunoglobulin. In contrast to T lymphocytes B cells do not require MHC-restricted processing of antigen for recognition, but the BCR can detect antigen in its particulate form. Nevertheless, most productive B cell responses require co-stimulation by cell–cell contact with T cells and/or antigen-presenting cells in the close proximity and/or via soluble mediators secreted by these cells.

Co-stimulation of B cell responses can further be mediated via pattern recognition receptors (PRR) expressed on B cells themselves. They provide the B lymphocyte with the capacity to sense microbial danger signals and potentially distinguish self from foreign antigen. This is most relevant considering the fact that B cells capture non-processed particulate antigens that contain the BCR-activating antigen in complexion with ligands for PRR.

Hans-Joachim Anders and Adriana Migliorini (eds.), *Innate DNA and RNA Recognition*, Methods in Molecular Biology, vol. 1169, DOI 10.1007/978-1-4939-0882-0_9, © Springer Science+Business Media New York 2014

Furthermore, concomitant stimulation of the BCR and PRR such as Toll-like receptors (TLR) can drive T cell-independent B cell responses. This type of response provides rapid antibody-mediated protection from pathogens and, most likely, only occurs under two physiological circumstances: (a) natural antibody production by innate immune B cells challenged with bacterial molecules, e.g., phosphocholine or pneumococcal polysaccharides and (b) antibody production upon secondary exposure of memory B cells to their cognate antigen. It should however be noted, that T cell help is a prerequisite for the activation of naïve B lymphocytes and for the de novo formation of antigen-specific memory B cells.

TLR are specialized receptors for the sensing of specific molecular targets typically found in microbial pathogens. To date, TLR represent the best studied pattern recognition receptors in B cell responses (reviewed in refs. 1, 2). Notably, TLR expression and functionality vary in B cell subsets and among the species. The most important difference is that responsiveness to TLR2 and TLR4 ligands is a constant feature on murine B cells, most prominently on B1 and marginal zone (MZ) B cell subsets, while the responsiveness of human B cells to the respective agonist is low (TLR2) to absent (TLR4) [2, 3]. By contrast, expression of intracellular TLR specialized in sensing of nucleic acids by TLR7 (RNA) and TLR9 (CpG DNA) is common to nearly all B cell subpopulations, and, thus, highlights the specific requirement for BCR-mediated antigen uptake in providing access for antigen-associated nucleic acids to TLR ligand as demonstrated in the following reference [4].

Synergistic BCR and TLR engagement drives T cell-independent B cell activation characteristic of certain disease states: (a) terminal differentiation of autoreactive B cells and autoantibody secretion is driven by BCR-mediated uptake of nucleic acid-bearing (auto-) antigens [5, 6] and (b) polyclonal B cell expansion is triggered by infecting pathogens expressing B cell superantigens [7, 8].

Thus, studying the role of nucleic acid sensing TLRs requires experimental systems that allow us to mimic and to control sequential BCR and TLR stimulation. Several approaches have been described that deliver TLR-stimulatory nucleic acids coupled to BCR stimuli to the endosome: the eldest experimental system was developed in the laboratory of Ann Marshak-Rothstein. Using a B cell line expressing a transgenic BCR specific for the Fc portion of IgG2a (AM14 B cells), these investigators showed that uptake of immune complexes consisting of anti-DNA antibodies and chromatin or anti-RNA antibodies with RNA resulted in TLR7/9-dependent B cell activation [5, 6]. Chaturvedi et al. described a protocol for the generation of anti-mouse IgM $F(ab')_2$ + phosphorothioate (PTO)-modified CpG DNA ODN bioconjugates and found that physical linkage of BCR and TLR9 ligand significantly augments B cell activation [4]. Similar results were obtained by the

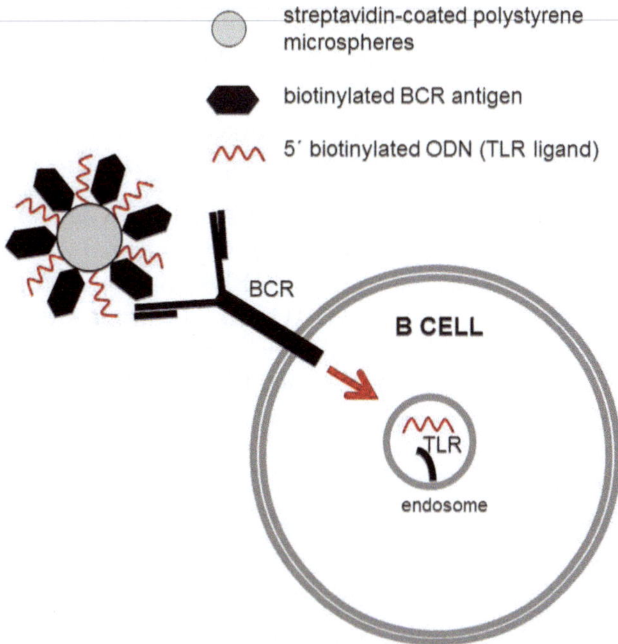

Fig. 1 Schematic drawing showing uptake of microspheres via coupled BCR antigens and subsequent stimulation of endosomal TLR via coupled TLR ligands, e.g., oligodeoxynucleotides (ODN). The protocol utilizes 5′ biotinylated unmodified CpG ODN (PO) because lack of the phosphorothioate modification makes BCR-mediated uptake of the ODN a prerequisite for stimulation of endosomal TLR. Consequently, B cell activation depends on BCR ligands coupled to microspheres and BCR specificity

group of Facundo Batista who studied the effect of combined BCR and TLR stimulation using beads coated with protein antigens (hen egg lysozyme, ovalbumin, and chicken γ-globulin) and/or CpG ODN as a TLR9 ligand, thereby targeting antigen-specific murine B cells [9].

Lack of transgenic BCR in human cells has remained an obstacle in studying human B cell activation. This chapter provides a protocol for concomitant stimulation of BCR and TLR in human B lymphocytes. Importantly, TLR activation is only initiated after endocytotic uptake of TLR-active ODN in a BCR-dependent manner as explained in Fig. 1.

2 Materials

2.1 CD19+ B Cell Isolation and Culture

1. 1.077 g/mL Ficoll Hypaque.
2. 0.9 % (w/v) Sodium chloride (NaCl).
3. Anti-human CD19 microbeads (Miltenyi Biotec, Bergisch-Gladbach, Germany).

4. Running Buffer (sterile, ready-to-use, Miltenyi Biotec) contains 1 % BSA, 2 mM EDTA, 0.09 % sodium azide.
5. Magnetic cell sorting device, i.e., AutoMACS™ (Miltenyi Biotec, Bergisch Gladbach, Germany).
6. 1× PBS without Ca^{2+} Mg^{2+}, pH = 7.2.
7. RPMI 1640 medium (GIBCO® Life Technologies, Darmstadt, Germany) supplemented with 3 mM L-Glutamine (GIBCO®), 0.01 M HEPES Buffer (PAA Laboratories, Pasching, Austria), 100 IU/mL Penicillin–Streptomycin (PAA Laboratories). Additionally 5–10 % ultralow IgG FCS (heat-inactivated at 56 °C for 1 h; GIBCO®) or autologous heat-inactivated serum is added.

2.2 Biotinylation of Ligands

1. Purified staphylococcal protein A (GE Healthcare, Uppsala, Sweden).
2. Tetanus toxoid (Statens Serum Institut, Copenhagen, Denmark).
3. 10 mM Sulfo-NHS-LC-Biotin solution (Thermo Scientific, Karlsruhe, Germany).
4. Zeba™ Desalt Spin Column (Thermo Scientific, Karlsruhe, Germany).
5. Components 3 and 4 contained in EZ-Link® Sulfo-NHS-LC-Biotinylation Kit (Thermo Scientific, Karlsruhe, Germany).
6. BCA Protein Assay (Thermo Scientific) or other standard method for determination of protein concentrations.

2.3 Coupling of Microspheres

1. Streptavidin-coated polystyrene microspheres (Bangs Laboratories, Fishers, IN, USA) with 0.13 μm diameter (non-fluorescent) or 0.39 μm diameter (fluorescent, dragon green).
2. Biotinylated $F(ab')_2$ goat anti-human IgM + IgA + IgD (Jackson ImmunoResearch, UK).
3. Biotinylated protein A or tetanus toxoid (see above, phosphorylation and coupling described below).
4. 5′-Biotinylated oligodeoxynucleotides (no phosphorothioate modification) (synthesized at MWG Eurofins, Ebersberg, Germany).

Sequences:

CpG 2006 PO: 5′→3′ TCGTCGTTTTGTCGTTTTGTCGTT

GpC 2006 PO: 5′→3′ TGCTGCTTTTGTGCTTTTGTGCTT

polyT: 5′→3′ TTTTTTTTTTTTTTTTTTTT

2.4 BrdU Incorporation Assay

1. Bromodeoxyuridine (BrdU) flow kit (BD Biosciences).
2. Flow cytometer.

2.5 ^{3}H-Thymidine Proliferation Assay

1. ^{3}H-methyl-thymidine solution (33 μCi/mL in RPMI 1640; PerkinElmer, Rodgau, Germany).
2. Filter mats and filter plates (PerkinElmer).
3. FilterMate Harvester (PerkinElmer).
4. Liquid scintillation solution (MicroScint™ 40, PerkinElmer).
5. Liquid scintillation counter (CHAMELEON™ V Liquid scintillation counter, Hidex, Turku, Finland).

2.6 Confocal Microscopy

1. DAPI (4′,6-Diamidino-2-Phenylindole, Dilactate; Molecular Probes, Leiden, The Netherlands) or Hoechst 33258 dye (Invitrogen, Karlsruhe, Germany).
2. 4 % Paraformaldehyde (PFA) (w/v) in 1× PBS.
3. Mounting Medium (Immunoselect Antifading Mounting Medium, Dianova, Hamburg, Germany).
4. Glass slides and cover slips (e.g., Marienfeld GmbH, Lauda-Königshofen, Germany).
5. Confocal Microscope.

3 Methods

3.1 CD19+ B Cell Isolation

1. PBMCs are purified from whole blood by density gradient centrifugation using Ficoll Hypaque (*see* Subheading 2.1, **item 1**). After washing isolation of CD19+ positive B cells is performed by using magnetically activated cell sorting (MACS) or alternative cell sorting method.
2. PBMCs are incubated with anti-human CD19 microbeads (Miltenyi Biotec) for 15 min at 4 °C according to the manufacturer's instructions.
3. The cells are washed and resuspended in 10 mL 1× PBS.
4. Positive selection is performed using a magnetic sorting device such as the AutoMACS™ (Miltenyi Biotec) using the programs "Possel_s" followed by "Possel."
5. The originating positive cell fraction is centrifuged, the supernatant discarded and purified cells are resuspended in medium (RPMI 1640 containing L-Glutamine, HEPES Buffer, and penicillin–streptomycin; *see* Subheading 2.1, **item 7**) at a concentration of 1×10^6 cells per 1 mL. The cells are stimulated directly or kept in an incubator (37 °C, 5 % CO_2, 95 % humidity) overnight.
6. Culture medium is supplemented with 5–10 % heat-inactivated autologous serum or low IgG FCS (*see* Subheading 2.1, **item 7**).
7. B cell purity is controlled by flow-cytometric analysis with fluorochrome-conjugated anti-CD20 antibody (BD Biosciences).

3.2 Biotinylation of Ligands

1. Protein A (SpA) and Tetanus toxin (TT) are biotinylated following the protocol suggested in the EZ-Link® Sulfo-NHS-LC-Biotinylation Kit (Thermo Scientific).
2. Immediately before use and according to instructions the 10 mM Sulfo-NHS-LC-Biotin solution is prepared by dissolving 2.5 mg Biotin in 450 μL ultrapure water or 1.1 mg in 200 μL, respectively.
3. An appropriate volume of biotin solution (depending on protein mass and concentration; equation provided by manufacturer) is added to the proteins and incubated for 1 h at RT.
4. Buffer exchange and removal of excess biotin are achieved using a Zeba™ Desalt Spin Column or other dialysis method.
5. The final protein concentrations are measured with the BCA protein Assay or other standard methods for determining protein concentrations.
6. Biotinylated protein solution is stored at −20 °C.

3.3 Coupling of Microspheres

1. Biotinylated ODNs (*see* **Note 1**) and proteins are coupled to streptavidin-coated polystyrene microspheres with a diameter of 0.13 μm (non-fluorescent microspheres used for stimulation) or 0.39 μm (fluorescent microspheres used for confocal analysis).
2. Per reaction 10 μL of microsphere suspension (~100 μg beads) is transferred to tubes and washed twice with 500 μL NaN_3 (10,000 × *g*, 5 min on a microcentrifuge).
3. Microspheres are resuspended in 100 μL 1× PBS.
4. Biotinylated stimuli are added at the concentrations indicated in Table 1 and incubated for 30 min at RT while shaking.
5. If two stimuli are to be conjugated to the same microspheres they are thoroughly mixed prior to incubation with microspheres.

Table 1
Concentration of biotinylated ligands used for coupling of microspheres

Stimulus (biotinylated)	Concentration of microsphere solution
Tetanus toxoid	4 μg/mL
Protein A	8 μg/mL
BCR ligation: $F(ab')_2$ anti-human IgM + IgA + IgG fragments (H + L)	4 μg/mL
CpG 2006 PO	2 μM

6. After incubation the microspheres are washed with 1× PBS and centrifuged at 4 °C as above.
7. The conjugates are resuspended in 200 μL PBS and stored at 4 °C.
8. For stimulation 10 μL of microsphere solution are used (*see* **Note 2**).
9. Uncoupled microspheres serve as control.

3.4 BrdU Proliferation Assay

1. Bromodeoxyuridine (BrdU), a synthetic analogue of thymidine, is incorporated into newly synthesized DNA of replicating cells, thereby substituting for thymidine. It can be detected and quantified with an anti-BrdU antibody after permeabilization of cells and denaturation of DNA.
2. Human CD19+ B cells are cultured in a 96-well U-bottom plate at 2×10^5 cells/well.
3. Cells are stimulated with 10 μL of different microsphere conjugates (0.13 μm diameter) and pulsed with 20 μL/well 1 mM BrdU solution.
4. On day 5 after stimulation cells are pelleted and washed.
5. Intranuclear BrdU staining is performed with fluorochrome-conjugated anti-BrdU antibody after fixation and permeabilization of cells and denaturation of DNA following the manufacturer's instructions (i.e., BrdU Flow Kit Staining Protocol).
6. Incorporated BrdU is quantified and analyzed by flow cytometry (*see* **Note 3**).

3.5 ^{3}H-Thymidine Incorporation Assay

1. CD19+ B cells are plated at $5–20 \times 10^4$ cells/well in a 96-well U-bottom plate and stimulated with 10 μL of different microsphere suspensions.
2. The cells are pulsed with 15 μL ^{3}H-methyl-thymidine solution (0.5 μCi/well) for 18 h.
3. After a total of 72 h of incubation, the cells are harvested onto filter mats and DNA-bound radioactivity quantified on a liquid scintillation counter (*see* **Note 3**).

3.6 Confocal Microscopy

1. 1×10^6 CD19+ B cells are stimulated with 20 μL fluorescent 0.39 μm beads-conjugates.
2. After 3 h cells are prepared for microscopy: cells are washed and resuspended in 1× PBS and transferred to wells of a 96-well plate or 1.5 mL eppendorf tubes.
3. For staining of nuclear DNA Hoechst (5 μg/mL) or DAPI (0.5 μM) are added to the cells in a 50 μL volume and incubated for 30 min at RT in the dark.

4. Cells are fixed in 4 %PFA/1× PBS for 10 min.
5. Alternatively, nuclear staining can be performed during fixation.
6. After staining and fixation cells are washed with PBS and resuspended in a small volume of glycerol-based mounting medium and applied to a glass slide, covered by coverslips and stored at 4 °C in the dark.
7. Analysis is performed on a confocal microscope using 63× or 100× objectives.

4 Notes

1. Due to their endosomal localization stimulation of nucleic acid-sensing TLR requires uptake of TLR ligands in a BCR-dependent manner. Here, we present a microsphere-based method for BCR-mediated internalization of TLR-active ODN. The data provided show data obtained using a TLR9-stimulatory ODN; the TLR9 ligand can also be substituted by immune stimulatory RNA.
2. Figure 2a shows that coupling of microspheres to BCR ligands is a prerequisite for binding of microspheres to B cells. We further demonstrate that induction of B cell proliferation is dependent on the presence of both the BCR and the TLR ligand (Fig. 2b, c) and, indeed, most efficient when a BCR stimulus and TLR ligand are physically associated, e.g., coupled to the same microsphere (Fig. 2b). Notably, induction of B cell proliferation occurs is a concentration-dependent manner (data not shown) and optimal concentrations of microspheres need to be determined for individual lots of coupled microspheres. Finally, we show that depending on the BCR stimulus this protocol can be used to trigger activation of specific B cell subsets as induced by protein A-bearing microspheres or expansion of antigen-specific B cells as achieved by stimulation with tetanus toxoid carrying microspheres (Fig. 2c).
3. It should be noted that using this experimental system we achieve T cell-independent induction of proliferation in human peripheral blood B cells. However, in contrast to stimulation with phosphorothioate (PTO)-modified CpG ODN we did not observe terminal differentiation of B cells or relevant immunoglobulin production in response to microsphere-based stimulation with TLR9 ligand; this observation could be due to reduced stability of TLR-stimulating nucleic acids when compared to PTO ODN.

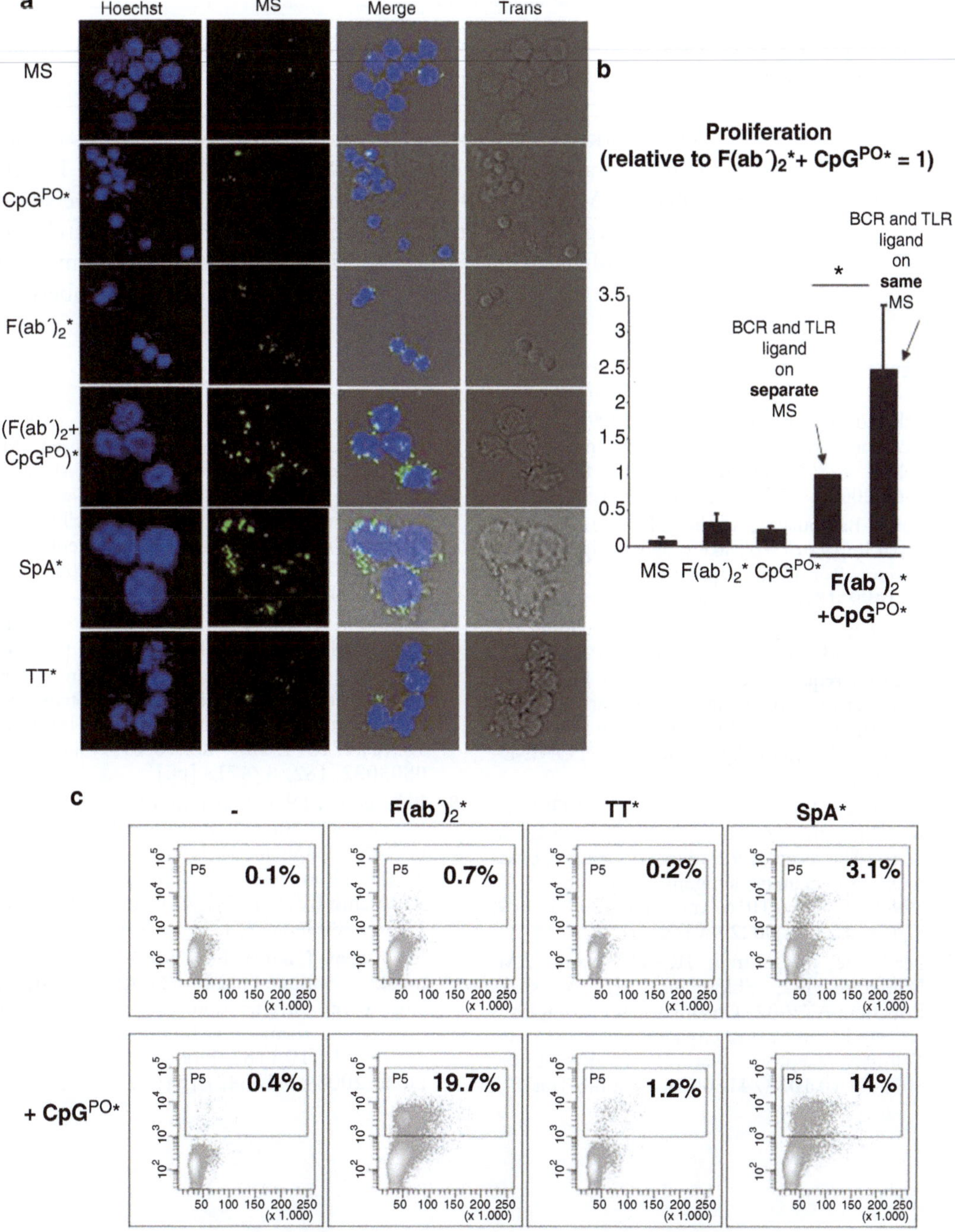

Fig. 2 Dual engagement of BCR and TLR9 using microspheres coupled with BCR and/or TLR ligands. Streptavidin-coated microspheres (MS) were coupled with or without biotinylated (*) BCR stimuli, e.g., anti-human Ig (F(ab′)$_2$*), protein A (SpA*), or tetanus toxoid (TT*) in the presence of absence of 5′ biotinylated oligodeoxynucleotides, e.g., TLR9 ligand CpG^{PO}*. Uncoupled microspheres were used as control. (**a**) CD19+ human peripheral blood B cells were incubated with green fluorescent microspheres (0.39 μm diameter). After 3 h the cells were washed, nuclei stained with Hoechst, and binding of MS analyzed by confocal microscopy. (**b**) CD19+ B cells were stimulated with non-fluorescent microspheres (0.13 μm diameter) and proliferation assessed by ^{3}H-thymidine incorporation. The diagram summarizes the results obtained in $n = 5$ independent experiments ± SEM. The results were normalized to F(ab′)$_2$* + CpG^{PO}* = 1 (=2,478 ± 1,424 cpm) to reduce the extent of donor variability. (**c**): B cells were stimulated with microspheres in the presence of BrdU. The cells were harvested on day 5 and proliferation quantified by the mean fluorescence intensities (MFI) of anti-BrdU staining. The graph depicts one representative experiment

Acknowledgements

This work was supported by the German Research Association (DFG) grants BE3841/2-1 and SFB 938 TP C to I.B.-D., the Olympia-Morata grant of the Medical faculty of the University of Heidelberg, Germany to I.B.-D. and the postgraduate research program "Differential activation and integration of signaling modules within the immune system" of the state of Baden-Württemberg.

References

1. Bekeredjian-Ding I, Jego G (2009) Toll-like receptors–sentries in the B-cell response. Immunology 128(3):311–323. doi:10.1111/j.1365-2567.2009.03173.x, IMM3173 [pii]
2. Meyer-Bahlburg A, Rawlings DJ (2012) Differential impact of Toll-like receptor signaling on distinct B cell subpopulations. Front Biosci 17:1499–1516, 4000 [pii]
3. Bekeredjian-Ding I, Foermer S, Kirschning CJ, Parcina M, Heeg K (2012) Poke weed mitogen requires Toll-like receptor ligands for proliferative activity in human and murine B lymphocytes. PLoS One 7(1):e29806. doi:10.1371/journal.pone.0029806, PONE-D-11-01385 [pii]
4. Chaturvedi A, Dorward D, Pierce SK (2008) The B cell receptor governs the subcellular location of Toll-like receptor 9 leading to hyperresponses to DNA-containing antigens. Immunity 28(6): 799–809. doi:10.1016/j.immuni. 2008.03.019, S1074-7613(08)00229-X [pii]
5. Leadbetter EA, Rifkin IR, Hohlbaum AM, Beaudette BC, Shlomchik MJ, Marshak-Rothstein A (2002) Chromatin-IgG complexes activate B cells by dual engagement of IgM and Toll-like receptors. Nature 416(6881):603–607. doi:10.1038/416603a, 416603a [pii]
6. Lau CM, Broughton C, Tabor AS, Akira S, Flavell RA, Mamula MJ, Christensen SR, Shlomchik MJ, Viglianti GA, Rifkin IR, Marshak-Rothstein A (2005) RNA-associated autoantigens activate B cells by combined B cell antigen receptor/Toll-like receptor 7 engagement. J Exp Med 202(9):1171–1177. doi:10.1084/jem.20050630, jem.20050630 [pii]
7. Jendholm J, Morgelin M, Perez Vidakovics ML, Carlsson M, Leffler H, Cardell LO, Riesbeck K (2009) Superantigen- and TLR-dependent activation of tonsillar B cells after receptor-mediated endocytosis. J Immunol 182(8):4713–4720. doi:10.4049/jimmunol.0803032, 182/8/4713 [pii]
8. Bekeredjian-Ding I, Inamura S, Giese T, Moll H, Endres S, Sing A, Zahringer U, Hartmann G (2007) Staphylococcus aureus protein A triggers T cell-independent B cell proliferation by sensitizing B cells for TLR2 ligands. J Immunol 178(5):2803–2812, 178/5/2803 [pii]
9. Eckl-Dorna J, Batista FD (2009) BCR-mediated uptake of antigen linked to TLR9 ligand stimulates B-cell proliferation and antigen-specific plasma cell formation. Blood 113(17):3969–3977. doi:10.1182/blood-2008-10-185421, blood-2008-10-185421 [pii]

Chapter 10

Mapping of Optimal CD8 T Cell Epitopes

Julia Roider, Thomas Vollbrecht, and Rika Draenert

Abstract

Defining the optimal epitope of a CD8 T cell response towards a certain antigen is a multistep procedure that requires the performance of peptide truncation design, ELISPOT peptide titration assays, and assessing the HLA class I restriction of the defined epitope via intracellular cytokine staining assays with B cell lines and epitope-specific CD8 T cell lines.

Key words Optimal epitope, CD8 T cell line (CTL), B cell line (BCL), Intracellular cytokine staining assay (ICS assay), HLA restriction, Peptide truncations, ELISPOT

1 Introduction

Methods for identifying physiologically relevant CD8 T cell epitopes are critically important not only for the development of vaccines but also for understanding the immune pathology of various diseases. As mapping an optimal CD8 T cell epitope is a multistep procedure that requires time, money, and patience, it is nowadays often replaced by bioinformatics tools [1].

Antigen-specific CD8 T cells following infection or immunization are typically assessed by measuring cytokine production after stimulation with overlapping peptides spanning the region of interest [2]. For fine mapping, peptide truncations of the epitope of interest are designed and their recognition is tested subsequently via ELISPOT titration assays. To define HLA class I restriction of the immunodominant epitope, intracellular cytokine staining assays and flow cytometry are performed afterwards, using epitope-specific CD8 T cell lines and B cell lines serving as antigen presenting cells in the assay. All of the applied methods have been described in the literature individually [2, 3, 5, 6]. Here, we have summarized all the procedures needed for the multistep process of experimentally mapping a CD8 T cell epitope. The epitope mapping starts when a CD8 T cell response towards a longer (10–20-mer) screening peptide has been found.

Hans-Joachim Anders and Adriana Migliorini (eds.), *Innate DNA and RNA Recognition*, Methods in Molecular Biology, vol. 1169, DOI 10.1007/978-1-4939-0882-0_10, © Springer Science+Business Media New York 2014

2 Materials

Prepare and store all materials at room temperature (unless stated otherwise).

2.1 ELISPOT Assays

1. Separate peripheral blood mononuclear cells (PBMCs) from whole blood according to the manufacturer's protocol (Ficoll density gradient). Incubate at 37 °C, 5 % CO_2 if not used immediately.
2. 96-well polyvinylidene difluoride-backed plates coated with 100 μL of anti-gamma-interferon (IFN-γ) mAb 1-D1k (1 μg/mL) and incubated overnight at 4 °C.
3. R10 medium: RPMI 1640 supplemented with 2 mM L-glutamine, 50 U/mL penicillin, 50 mg/mL streptomycin, 10 mM Hepes, and 10 % heat-inactivated fetal calf serum (FCS). Store at 4 °C and warm to 37 °C before use.
4. Reagents:
   - Phosphate-buffered saline (PBS), phytohemagglutinin (PHA), wash solution: PBS with 1 % heat-inactivated FCS. Store at 4 °C (*see* **Note 1**).
   - Antibody mAb 7B6-1-Biotin (0.5 μg/mL), 1:20,000-diluted streptavidin–alkaline phosphatase conjugate. Store at 4 °C.
   - Color solution: 12 mL aqua dest. + 500 μL Tris–HCl buffer 2.5 mol/L pH 9.7 + each 125 μL color reagent A and B (Bio-Rad Laboratories, Hercules, USA). Prepare two color solutions, one for each reagent. Store color reagents at −20 °C and prepare solutions fresh for each use.
   - 0.05 % Tween 20.
5. AID ELISPOT Reader.
6. Synthetic peptides corresponding to the antigen in question at 200 μg/mL. Store at 4 °C.

2.2 Antigen-Specific CD8 T Cell Lines

1. PBMCs of individual in question (seven million). Incubate at 37 °C, 5 % CO_2 if not used immediately.
2. Feeder cells: Irradiated (3,000 Rad) allogeneic PBMCs of irrelevant donor (25 million). Incubate at 37 °C, 5 % CO_2 if necessary.
3. Cell culture flasks in different sizes.
4. R10 medium and R10/IL2 medium: RPMI 1640 (PAA) supplemented with 2 mM L-glutamine, 50 U/mL penicillin, 50 mg/mL streptomycin, 10 mM Hepes, 10 % heat-inactivated FCS, and 100 U of recombinant interleukin-2 per mL. Store at 4 °C and warm before use to 37 °C.
5. Synthetic peptides at 200 μg/mL. Store at 4 °C.

2.3 EBV-Transformed B Lymphoblastoid Cell Lines

1. 24-well plates.
2. Epstein-Barr virus stock is generated from B95-8, an EBV producing cell line. B95-8 cells are thawed, placed in T75 flask and fed with 15 mL R10 medium containing 0.4 μg/mL Phorbol 12-myristate 13-acetate (PMA). After 11–14 days, spin cells down and filter supernatant with 0.2 μm-filter and freeze as EBV-transforming supernatant in several aliquots at −80 °C (*see* **Note 2**).
3. R20 medium: RPMI 1640 supplemented with 2 mM L-glutamine, 50 U/mL penicillin, 50 mg/mL streptomycin, 10 mM Hepes, and 20 % heat-inactivated FCS. Store at 4 °C and warm to 37 °C before use.
4. Reagent: Cyclosporin A at 0.4 μg/mL. Store at −20 °C (*see* **Note 3**).

2.4 Intracellular Cytokine Staining Assays

1. 24-well plates, FACS tubes.
2. R10 medium. Store at 4 °C and warm to 37 °C before use.
3. Reagents: Brefeldin A, store at −20 °C; Fix/Perm Solution A and B.
4. Antibodies: anti-CD8-PerCP, anti-CD4-APC and anti-IFN-gamma-FITC, co-stimulants: anti-CD28, anti-CD49d (Becton Dickinson). Store at 4 °C.
5. FACS wash solution: PBS with 1 % heat-inactivated FCS. Store at 4 °C.
6. Synthetic peptides at 200 μg/mL. Store at 4 °C.
7. FACSCalibur™ flow cytometer, CellQuest™ and FlowJo software.

2.5 Intracellular Cytokine Staining for HLA Class I Restriction

1. As described in Subheading 2.4.
2. DNA extraction for HLA class I is performed according to manufacturer's protocol of the chosen Kit.

3 Methods

Carry out all procedures at room temperature unless stated otherwise. If not stated otherwise, washing is performed at $352 \times g \times 10$ min at 20 °C.

3.1 ELISPOT Peptide Truncation and Titration Assay

1. Performance of ELISPOT assay as described [2]. Screening is usually performed with 10–20-mer peptides that overlap by ten amino acids. Final peptide concentration is 12.5 μg/mL and PBMCs are added at $0.5–1 \times 10^5$ per well.
2. Knowing HLA class I alleles of the subject of interest is essential for the following steps and HLA typing should be initiated

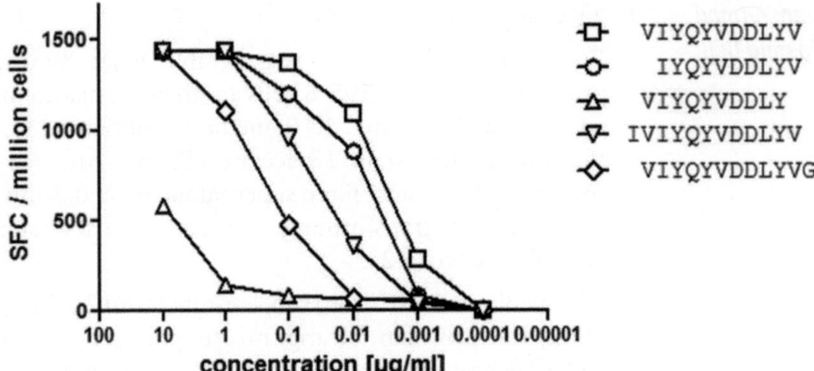

Fig. 1 ELISPOT titrations of truncated peptides for the epitope VIYQYVDDLYV. *X*-axis: peptide concentration in increasing dilution. *Y*-axis: magnitude of CD8 T cell responses expressed in spot forming units per million PBMCs. Screening peptide was RKQNPDIVIYQYVDDLYV. In this example, you can observe that the optimal epitope is the 11-mer VV11 (*squares*). To fine-map the epitope, peptide truncations adding/deleting one amino acid at the N- and the C-terminal position have been used

as soon as possible. For assessing the position of an optimal epitope (usually 8–11-mers for CD8 T cells) within the screening peptide (10–20-mers), CD8 T cell recognition of neighboring peptides has to be considered, as they often overlap by ten amino acids. To narrow down the quantity of possible epitopes within the screening peptide, bioinformatic tools (e.g., http://www.syfpeithi.de) can be consulted or literature searched for already described optimal epitopes. If this approach does narrow the possibilities, the screening peptide can be divided in half for the first test. If the epitope is more or less defined, truncations of the peptide are designed by creating variants of the suspected optimal epitope by adding or deleting one amino acid to or from the N- or C-terminal residues of the sequence of interest [3]. Please *see* Fig. 1 [4] for a graphic example from our laboratory. Peptide comparisons are done in double and in ≥2 independent experiments.

3. For peptide titration assays, peptides are used at concentrations of 12.5 μg/mL–1.25×10^{-4} μg/mL using tenfold dilutions. Each peptide comparison is done in double and peptide comparisons are done in ≥2 independent experiments.
4. Gamma interferon-producing cells are counted by direct visualization on an AID ELISPOT Reader and are expressed as spot-forming cells (SFC) per 10^6 PBMCs. Negative controls are ≤5 spots. Wells were counted as positive if they are ≥50 SFC/10^6 PBMCs (*see* **Notes 4–6**).

3.2 Antigen-Specific CD8 T Cell Lines [3]

1. Resuspend PBMCs in R10 medium.
2. Set up lines in a T25 flask:

- Two million PBMCs of individual of whom epitope should be mapped in 5 mL R10.
- 25 million irradiated feeder cells of irrelevant donor in 10 mL R10 (*see* **Note 7**).
- 20 mL R10/IL-2 medium (*see* **Note 8**).
- Five million peptide-pulsed antigen-presenting cells in 1 mL R10, either autologous PBMCs or autologous B cell line (*see* Subheading 3.3. for further details). Prepare as follows:
  - Add peptide to antigen-presenting cells. Final peptide concentration should be 10 μg/mL.
  - Incubate at 37 °C, 5 % CO_2 for 2 h (*see* **Note 9**).
  - Irradiate at 3,000 Rad (*see* **Note 10**).
  - Wash three times in R10 to remove loose peptide and resuspend in 1 mL R10.

3. Feed for the first time on day 4 by exchanging about 7 mL R10/IL2 medium. Afterwards, feed twice a week (*see* **Notes 11–12**).
4. After 7–10 days, test for specificity as described in Subheading 3.4.
5. After 14 days, antigen-specific CD8 T cell lines should be restimulated with peptide-pulsed antigen presenting cells as described above (*see* **Note 13**).

3.3 EBV-Transformed B Lymphoblastoid Cell Lines [5]

1. Resuspend PBMCs of individual in question (e.g., carrying a certain HLA allele) in R20 medium at a final concentration of 10 million/mL (*see* **Notes 14–15**).
2. Set up lines on 24-well plate with a total volume of 1.5 mL per well
   - Ten million PBMCs in 1 mL R20 (ten million cells per well) (*see* **Note 16**).
   - 100 μL EBV supernatant (*see* **Note 17**).
   - 30 μL Cyclosporin A stock at 0.4 μg/mL (*see* **Note 18**).
   - 370 μL R20.
3. Incubation is performed at 37 °C, 5 % CO_2.
4. First time feed after 2 days by adding 0.5 mL R20 per well (*see* **Note 19**).
5. After that, feed twice a week for next 6–7 weeks by exchanging 1 mL of medium per well (e.g., after 5, 3, 4, 3, 4, 3, 4, … days). It usually takes 8–9 weeks for cell transformation (*see* **Note 20**).
6. After 3–4 weeks cells start to grow—expand them to one more well. Continue feeding twice a week.
7. After 8–9 weeks, transfer cells to a T25 flask (*see* **Note 21**). Again keep feeding twice a week.

3.4 Intracellular Cytokine Staining for CD8 T Cell Responses as Described in [6] with Minor Modifications

1. PBMCs are thawed and rested in R10 medium overnight at 37 °C, 5 % CO_2 (*see* **Note 22**).
2. Resuspend in 1 mL R10 in one well of 24-well plate at $0.5–2 \times 10^6$/mL.
3. For stimulation, add 4 μM synthetic peptide
   - For negative control no peptide is added (*see* **Note 23**).
4. Add anti-CD28 and anti-CD49d at 1 μg/mL per well (*see* **Note 24**).
5. Incubate at 37 °C, 5 % CO_2 for 1 h.
6. Add 10 μL of 1 mg/mL Brefeldin A.
7. After 5 h of incubation at 37 °C, 5 % CO_2, cells are washed with PBS/1%FCS at 4 °C and stained with surface antibodies anti-CD8-PerCP (1 μL/mL), anti-CD4-APC (4 μL/mL).
8. Incubate at 4 °C for 30 min.
9. After washing at 4 °C, 100 μL of Fix/Perm Solution A is added.
10. Incubate at room temperature in the dark for 15 min (*see* **Note 25**).
11. After washing at 4 °C, 100 μL of Fix/Perm Solution B is added.
12. Incubate at room temperature in the dark for 15 min.
13. Add intracellular antibodies (anti-IFN-gamma-FITC 15 μL/mL).
14. Incubate at 4 °C for 30 min.
15. Cells are then washed at 4 °C and analyzed using a FACSCalibur™ flow cytometer, CellQuest™ and FlowJo software. An IFN-gamma response is considered positive if threefold above background of negative control.

3.5 Intracellular Cytokine Staining for HLA Class I Restriction as Described in [2] with Minor Modifications

The assay is performed to determine the HLA class I-restriction of a defined optimal epitope. The protocol has the same principle as described in Subheading 3.4; the major difference is that peptide-specific CD8 T cell lines are not stimulated with peptide, but with partly HLA-matched peptide-pulsed B cell lines.

1. Choose B cell lines according to study subjects' HLA class I alleles (e.g., HLA-A*02, HLA-A*03, HLA-B*15, HLA-B*57, HLA-Cw*03, HLA-Cw*07). In an ideal approach, for each HLA allele of the study subject there is one B cell line that expresses just one HLA allele of the study subject in question. So there are six B cell lines (for the six possible different HLA alleles) that do not share a second HLA allele with the study subject. However, in many cases this is not possible, i.e., a B cell line shares more than allele with the study subject.

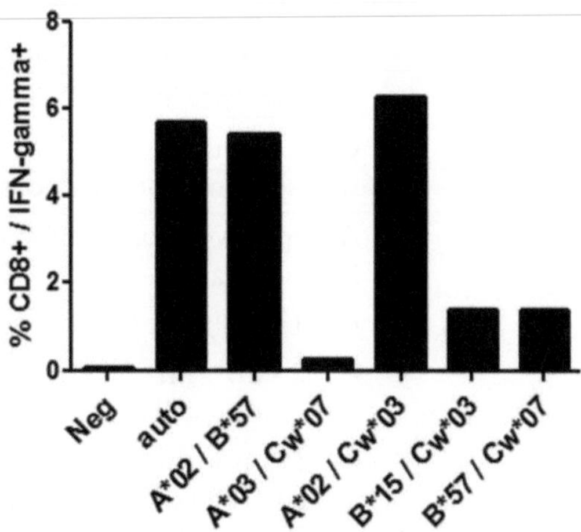

Fig. 2 HLA restriction of CD8 T cell responses of the epitope VIYQYVDDLYV. HLA class I alleles of study subject: HLA-A*02:01, -A*03:01; -B*15:01, -B*57:03; -C*03:04, -C*07:01. *X*-axis: negative control (peptide-specific CD8 T cell line without peptide), peptide-specific CD8 T cell line with peptide (positive control), B cell lines with peptide. For the B cell lines, only the matching HLA alleles are mentioned. *Y*-axis: interferon-gamma production as measured by flow cytometry. In this example you can clearly define that the epitope is HLA-A*02 restricted. Although it was not possible to find a B cell line that matched only the HLA-A*02:01 allele (*third bar*), it can be excluded that the epitope is restricted by HLA-B57:03 by using a second B cell line expressing the B*57 allele (*seventh bar*). The same applies to the Cw*3 allele

In this case, B cell line combinations need to be found that lead to an exclusion of certain HLA alleles. Please *see* Fig. 2 [4] for graphic explanation. B cell lines should be in culture for 7 days before performing the assay. Please *see* Subheading 3.3 for details.

2. Peptide-specific CD8 T cell lines are washed and rested in R10 overnight (*see* **Note 26**). Final concentration for the assay is 1×10^6/mL. Amount of CD8 T cell lines needed per assay:
   - Two million of peptide-specific CD8 T cell lines serving as positive autologous control (with peptide) and internal background control (without peptide) for the whole assay (*see* **Note 27**).
   - For each B cell line, two million of peptide-specific CD8 T cell lines (1×10^6 with peptide-loaded B cell line and 1×10^6 with B cell line without peptide for negative control).
3. Next day, bring B cell lines to a final concentration of 1×10^6/mL. Amount of each B cell line needed per assay:
   - 2×10^5 B cells stimulated with peptide (antigen presenting cells).
   - 2×10^5/mL B cells without peptide (*see* **Note 28**).

4. Add 1 mM of synthetic peptide to 2×10^5 B cells (per B cell line) and to 1×10^6 CD8 T cells (positive autologous control) and incubate at 37 °C, 5 % CO_2 for 1 h.
5. Cells (including 2×10^5 B cells per B cell line without peptide stimulation as negative control) are washed three times.
6. Add per well (one well corresponding to one FACS tube) of 24-well plate:
   - 1×10^6/mL CD8 T cells (effector cells).
   - 2×10^5/mL B cells stimulated with peptide (antigen presenting cells) *OR* 2×10^5/mL B cells without peptide.
   - Do not forget the controls per assay: 1×10^6 peptide-specific CD8 T cells with peptide (positive autologous control) and 1×10^6 peptide-specific CD8 T cells without peptide (background control). Here, no B cells are added.
7. Incubate at 37 °C, 5 % CO_2 for 1 h.
8. Follow protocol "intracellular cytokine staining for CD8 T cell responses" starting at **step 4**.

4 Notes

1. PHA stimulates nonspecific lymphocyte interferon-gamma production in cell culture and thus serves as a positive control for ELISPOT assays.
2. Check supernatant for mycoplasma by PCR methods.
3. Cyclosporin A is a hydrophobic substance, its solubility is an important technical point. For creating a stock, dissolve 1 mg Cyclosporin A in 0.1 mL of absolute ethanol. Add 0.02 mL Tween 80 and mix solution well. Add RPMI 1640 medium drop-wise until the final volume reaches 1.0 mL (stock concentration 1 mg/mL).
4. When the negative control turns out positive ($\geq$50 SFC/10^6 PBMCs), it can be a sign of unspecific activation of the cells. This is observed more frequently in CD8 T cell lines than in PBMCs. Therefore, CD8 T cell lines have to be washed twice and placed in R10 medium without IL-2 overnight to reduce background.
5. If the negative control turns out positive ($\geq$50 SFC/10^6 PBMCs), wells are only counted positive if number of SFC/10^6 PBMCs are more than three times greater than the negative control.
6. If the response is weak, consider adding more cells per well, e.g., 2×10^5. If the response is too high (too many spot-forming cells), consider adding less cells, e.g., 0.2×10^5. This can be observed in highly specific CD8 T cell lines.

7. Consider adding more feeder cells (e.g., 30 million).
8. Use freshly prepared medium for generation of cell lines.
9. During incubation of PBMCs with peptide-pulsed antigen presenting cells, flip the tube gently twice to get cells in better contact.
10. Consider a longer irradiation time to inhibit unspecific overgrow of feeder cells.
11. When feeding, remove flasks carefully from the incubator to avoid disturbing the CD8 T cells growing on the bottom.
12. Try to keep your cell lines in an incubator separate from cultures that need to be removed often. Keeping your cell lines in a separate incubator will also minimize their temperature and CO_2 discomfort. Try to minimize time outside of preferred conditions.
13. You have to be aware that the success of an antigen-specific CD8 T cell line depends on the immunological capability of the individual of interest. In the case of an impaired immune response (e.g., HIV infection), it can be difficult to establish an antigen-specific CD8 T cell line.
14. Use freshly prepared medium for generation of cell lines.
15. If the PBMC donor has an impaired immune response (e.g., HIV), B cell transformation might not work as well.
16. You can vary PBMC concentration from 5 to 15 million PBMCs per well, Cyclosporin A from 0.2 to 1 μg/mL and EBV supernatant from 75 to 500 μL (final volume per well should not exceed 1.5 mL), for us the above given concentrations worked the best.
17. Consider breeding fresh EBV virus, it works better than frozen stock.
18. Cyclosporin A inhibits T cells. Discard remaining Cyclosporin A and EBV stock in the aliquots—do not refreeze.
19. Wait until the first exchange of medium, you might take out too much of the EBV.
20. When you exchange medium aspirate wells gently—do not disturb cells at the bottom. When cells start to grow, they rest at the bottom of the well and media will change color to yellow. Colony forming of cells (clumps) can be visualized macroscopically.
21. Exchange flasks every 2 months for avoiding contamination.
22. The assay works reliably down to 100,000 cells.
23. As a positive control, PHA can be used, but as the assay works well with peptides, we normally do not need it.

24. When performing a regular protocol for specificity of CD8 T cell responses, we do not add anti-CD28, anti-CD49d, as they are costimulants and the assay works well with just peptide.
25. It is possible to interrupt the protocol after **step 10** and keep the FACS tubes overnight at 4 °C.
26. HLA-restriction assays reliably work if your CD8 T cell line is specific of the corresponding peptide. Antigen specificity of used CD8 T cell line does not necessarily have to be high. The assay worked for us starting from a specificity >0.35 %.
27. Peptide-specific CD8 T cell lines itself produce a certain interferon-gamma background. This is the reason why you need peptide-specific CD8 T cells without peptide as a background control.
28. Cells of distinct individuals (as always in this assay) do react with each other and produce a certain interferon-gamma background. This is the reason why you compulsorily need a negative control (B cell line without peptide) with every B cell line.

References

1. Martin W, Sbai H, De Groot AS (2003) Bioinformatics tools for identifying class I-restricted epitopes. Methods 3:289–298
2. Altfeld MA, Trocha A, Eldridge RL et al (2000) Identification of dominant optimal HLA-B60- and HLA-B61-restricted cytotoxic T-lymphocyte (CTL) epitopes: rapid characterization of CTL responses by enzyme-linked immunospot assay. J Virol 74:8541–8549
3. Walker BD, Flexner C, Birch-Limberger K et al (1989) Long-term culture and fine specificity of human cytotoxic T-lymphocyte clones reactive with human immunodeficiency virus type 1. Proc Natl Acad Sci U S A 86(23): 9514–9518
4. Vollbrecht T, Eberle J, Roider J et al (2012) Control of M184V HIV-1 mutants by CD8 T-cell responses. Med Microbiol Immunol 201(2): 201–211
5. Walker BD, Chakrabarti S, Moss B et al (1987) HIV-specific cytotoxic T lymphocytes in seropositive individuals. Nature 328(6128):345–348
6. Draenert R, Altfeld M, Brander C et al (2003) Comparison of overlapping peptide sets for detection of antiviral CD8 and CD4 T cell responses. J Immunol Methods 275(1–2):19–29

Chapter 11

A Modular Approach to Suppression Assays: TLR Ligands, Conditioned Medium, and Viral Infection

Viktor H. Koelzer and David Anz

Abstract

The suppressive function of regulatory T-cells (Treg) requires precise control to allow an efficient adaption of the T-cell response to the requirements of the immune defense. In the setting of infection, an abrogation of the suppressive effect of Treg on the activation and proliferation of T-effector (Teff) cells is a central precondition to allow fast and efficient clearance of the infectious agent. Experimentally, the suppressive function of Treg on Teff can be indirectly measured in coculture proliferation assays. This versatile tool provides a readout of T cell proliferation in the presence of Treg through the measurement of a proliferation marker such as the incorporation of radioactively labeled thymidine (^{3}H Thymidine), carboxyfluorescein succinimidyl ester (CFSE) or 5-Bromo-2′-deoxyuridine (BrdU). In a modular approach, the culture conditions can thereby be adapted to evaluate the effect of any cell type, live and inactivated microorganisms, molecularly defined immunostimulatory ligands, and cytokines on the interplay of Teff and Treg function. Here, we demonstrate how the suppression assay can be used as a multifunctional tool to provide insights into the interaction of Treg with Teff under a variety of conditions in vitro.

Key words Suppression assay, Conditioned medium, Viral infection, BrdU, Regulatory T-cells, Functional assay

1 Introduction

Regulatory T-cells (Treg) are central to the maintenance of immunological homeostasis and exert active and dominant control over autoreactive T-cells. This dominant suppressive effect of Treg on Teff has been attributed to the competitive consumption of IL-2 and the transfer of inhibitory cAMP to Teff [1]. Further, Treg can inhibit the activation and differentiation of dendritic cells (DC) through the expression of IL-10 and TGF-β [2] and mediate the differentiation of Teff into Treg, a process which has recently also been attributed to the effects of IL-35 [3].

Treg of thymic origin (also called natural Treg) are CD4+ T cells that are characterized by the constitutive expression of CD25, the α-chain of the high affinity heterotrimeric IL-2 receptor, and

Hans-Joachim Anders and Adriana Migliorini (eds.), *Innate DNA and RNA Recognition*, Methods in Molecular Biology, vol. 1169, DOI 10.1007/978-1-4939-0882-0_11, © Springer Science+Business Media New York 2014

the transcription-factor Foxp3 [4]. Foxp3 is a DNA-binding forkhead/winged helix transcription factor, which determines the development, phenotypic features, and function of Treg [5]. Studies in Foxp3 GFP knock-in murine models provided conclusive evidence that no other cell population expresses Foxp3 in the mouse, while CD25 is also expressed on activated CD4+ Teff [6]. Consequently, Foxp3+ Treg represent about 70 % of murine CD4+ CD25+ T cells, a fact that needs to be kept in mind in functional experiments relying on CD25 as a selection marker.

Following T-cell receptor (TCR) activation, Treg are capable of effectively suppressing the activation, proliferation and function of Teff [7]. Interestingly, the effective antigen concentration needed for the activation of Treg is many times lower than for the activation of Teff [8]. This lowers the activation threshold for Treg in comparison to autoreactive Teff and causes an effective early activation of Treg in the presence of MHC-I bound autoantigen. At the same time, TCR stimuli can be used to activate Treg in culture and to measure their effect on the proliferation of the Teff responder population in vitro [9].

It is essential that the suppressive function of Treg can be adapted to the requirements of the immune defense through the effects of microbial danger signals [10]. In the case of infection, the suppression of the T-cell response by Treg has to be transiently abrogated to allow the generation of an efficient T-cell response against invading pathogens. Indeed, several research groups have recently demonstrated that immune stimulation through microbial danger signals that are recognized through a group of evolutionary highly conserved pattern recognition receptors (Toll-like receptors, TLR) can cause a complete loss of Treg function [9, 11]. Especially indirect effects of TLR-ligands on Treg were demonstrated using suppression assays containing Treg, Teff, and DC. DC can recognize the presence of the bacterial PAMP lipopolysaccharide (LPS) and demethylated Cytidine-phosphatidyl-Guanosine DNA-dinucleotides (CpG ODN) through TLR4 and TLR9 respectively [9]. This leads to the activation of nuclear factor (NF)-kappa B (NFκB) over the adapter-protein myeloid differentiation factor 88 (MyD88) [12]. NFκB initiates the transcription of proinflammatory cytokines such as IL-6, IL-12, and tumor necrosis factor-α (TNF-α) which are secreted into the medium of the coculture [13]. Using conditioned medium from TLR-activated DC for the stimulation of Treg in suppression assays, a joint IL-6 dependent mechanism of the indirect influence of ligands for TLR4 and TLR9 on Treg was demonstrated [9]. These findings have recently been extended for immunostimulatory RNA-oligonucleotides (RNA-ORN), which specifically stimulate the TLR7 on dendritic cells [14].

These insights into the indirect effects of TLR-activated DC on Treg function have recently been expanded by several studies demonstrating a direct effect of bacterial TLR-agonists on Treg. Murine T-cells express TLR1, 2 and TLR6, Treg additionally selectively express TLR4, 5, 7, and 8 [11, 15]. Using suppression assays, a direct abrogation of Treg function by stimulation of TLR2 and TLR8 on Treg themselves was demonstrated [16, 17].

For viral infection, both direct and indirect effects on Treg function have been described [11, 18]. Interestingly, active EMCV virus can be recognized by Treg through the expression of melanoma differentiation-associated gene 5 (MDA-5) and causes loss of Treg function in suppression assays [11]. Using a novel FACS based readout for incorporation of BrdU in the cell culture, the proliferation in each of the cellular components within the suppression assay after viral infection could be measured, indicating that the effect of viral infection was indeed due to disinhibition of the Teff population rather than proliferation of both Treg and Teff [11].

In summary, suppression assays are a versatile tool, which allow the assessment of Treg function in a variety of experimental settings. Starting with the basic interaction of just Treg and Teff, a variety of stimuli can be used for TCR-activation, both in presence or absence of different types of DC. Further, the effect of inflammatory stimuli on the interplay of Treg to Teff can be investigated, ranging from molecularly defined TLR ligands and immunostimulatory oligonucleotides to cytokines in conditioned medium and active viruses. In the following, we will illustrate a modular and systematic approach to achieve the optimum results for this large variety of experimental settings.

2 Materials

2.1 Isolation of Treg, Teff and DC

1. Mouse spleen or lymph nodes (*see* **Note 1**).
2. 40 μm cell strainer.
3. 1× PBS without $Ca^{2+}Mg^{2+}$, pH = 7.2.
4. Ortho-mune erythrocyte lysis buffer: diluted at 1:200 in PBS; pH 7.4.
5. MACS® Buffer: PBS, 10 Vol % FCS, 2 mM EDTA; pH 7.2.
6. For isolation of Treg and Teff: Regulatory T-Cell Isolation Kit, murine; For isolation of CD11c+ DC: CD11c+ Dendritic Cell Isolation Kit, murine.
7. Magnetic cell sorting device, i.e., QuadroMACS™ and Mini MACS™ (Miltenyi Biotec).
8. T-cell culture medium: RPMI 1640 VLE, 10 Vol % FCS, 100 IU/mL penicillin, 100 μg/mL streptomycin, 2 mM glutamine, 1 Vol % sodium pyruvate, 1 Vol % nonessential amino acids, 0.0001 Vol % 2-mercaptoethanol; pH 7.4.

Table 1
TCR-stimuli for use in proliferation assays

Name	Specificity	Clone	Isotype	Concentration
Anti-CD3 antibody: Soluble stimulatory antibody to induce T-cell proliferation in presence of DC.	CD3ε, mouse	500A2	Syrian Hamster IgG2, k	0.1 μg/mL
Anti-CD3 T-cell stimulation beads: MicroBeads covalently bound with anti-CD3 antibody. Induce proliferation with or without addition of DC.	CD3, mouse	145-2C11	Armenian Hamster IgG1, k	1:10 (Bead–Cell ratio)
Anti-CD3-CD28 T-Cell expander beads: Commercially available MicroBeads coated with anti-CD3 and anti-CD28 antibodies	CD3, mouse CD28, mouse	No information available	No information available	1:50–1:100 (Bead–Cell ratio)

2.2 DC-Dependent Suppression Assay

1. Freshly isolated Treg, Teff, DC (*see* **Note 2**).
2. T-cell culture medium (*see* Subheading 2.1).
3. TCR Stimulus (*see* Table 1).
4. TLR-Stimulus (control and experimental, *see* Table 2).
5. Sterile 96-well polystyrene flat bottom cell culture plates, light-impermeable.

2.3 BrdU Incorporation Assay (Chemoluminescence)

1. Bromodeoxyuridine (BrdU), buffers for fixation, permeabilization, and denaturation, and peroxidase-conjugated anti-BrdU antibody can be purchased as BrdU Cell Proliferation Chemoluminescence ELISA kit (Roche Diagnostics, Mannheim, Germany).
2. Microplate luminometer.
3. Sterile 96-well polystyrene flat bottom cell culture plates, light-impermeable.

2.4 BrdU Incorporation Assay (FACS)

1. FACS-Buffer: PBS, 2 % FCS.
2. FACSFlow and FACSSafe.
3. FOXP3 Fix/Perm Buffer and FOXP3 Perm Buffer for intracellular staining (both from eBioscience, Frankfurt, Germany).
4. 0.05 mg/mL DNAse I in PBS.
5. Paraformaldehyde (PFA) Solution: PBS, 2 % PFA.
6. FACS antibodies to CD4, Foxp3, and BrdU (*see* Table 3).

Table 2
Overview of TLR ligands for application in suppression assays

Name	Receptor	Description	Concentration (µg/mL)	Reference
Pam_3CSK_4	TLR2	Lipohexapeptide Pam 3-Cys-Ser-$(Lys)_4$	2	[17]
poly(I:C)	TLR3	Long DS RNA 5′-ICIC … ICIC-3′	2–10	[Unpublished data]
LPS	TLR4	Lipopolysaccharide of gram-negative bacteria	1	[9]
Flagellin	TLR5	Protein-component of bacterial flagella	1–5	[21]
9.2dr RNA PTO DOTAP[a]	TLR7	Single-stranded phosphorothioate modified RNA 5′-UGU CCU UCA AUG UCC UUC AA-3′	1–2	[11]
CL097	TLR7/8	*Small molecule*, imidazoquinoline	0.2–0.04	[11]
R848	TLR7/8	*Small molecule*, imidazoquinoline	0.2–0.04	[11]
Imiquimod	TLR7	*Small molecule*, imidazoquinoline-amine	0.2–1	[11]
CpG A 2216	TLR9	CpG-ODN 2216 5′-GGG GGA CGA CG TCG GGG GG-3′	0.6–3	[Unpublished data]
CpG B 1826	TLR9	CpG-ODN 1826 5′-CCA TGAC GTT CCT GAC GTT-3′	0.6–3	[11]

[a]A polyA PTO of the same length can be used as a control for the sequence specific recognition of RNA 9.2.dr by TLR7

Table 3
FACS antibodies

Specificity	Clone	Species	Isotype	Label	Concentration
CD4	GK1.5	Rat	IgG_{2b}, κ	FITC	1:100
CD11b	M1/70	Rat	IgG_{2b}, κ	PerCP	1:75
CD25	PC61	Rat	IgG_1, λ	PE	1:75
Foxp3	FJK-16s	Rat	IgG_{2a}	APC	1:75
BrdU	Bu20a	Mouse	IgG1, κ	PE	1:100

2.5 Conditioned Medium

1. Freshly isolated splenocytes or sorted CD11c+ DC.
2. T-cell culture medium (*see* Subheading 2.1).
3. TLR-stimulus (*see* Table 2).
4. Sterile 96-well polystyrene flat bottom cell culture plates.

2.6 Viral Infection

1. Viral particles in stock solution. Infection is best carried out at 1 MOI for EMCV, Sendai virus, and VSV [11].
2. FCS-free medium (RPMI 1640 VLE) for incubation.
3. Chlorine for inactivation of infectious particles and contaminated medium.

3 Methods

3.1 Isolation of Treg and Teff

1. Murine spleens are mechanically disrupted using a 40 μm cell strainer; cells are then washed in cold PBS followed by lysis of erythrocytes. After washing, the cells are resuspended in MACS® buffer. Treg and Teff are then sorted in a two step-procedure using magnetically activated cell sorting (MACS®, please refer to manufacturer's instructions for further detail) or an alternative cell sorting method. When using the MACS® procedure, a negative selection for CD4+ cells from whole splenocytes is performed in a first step (*see* **Note 3**). This is followed by positive selection of CD4+CD25+ regulatory T cells. The residual CD4+, CD25− Teff population represents an ideal responder population for the use in suppression assays. At the same time, the CD4−T cell depleted splenocytes can be used as an APC fraction in suppression assays or for the generation of conditioned medium and should not be discarded.
2. After isolation, Teff and Treg are resuspended in T-cell medium and can be directly used for the suppression assay (*see* **Note 4**). Alternatively, both Treg and Teff can be taken into culture at

37 °C, 5 % CO_2, 95 % humidity using a concentration of 1×10^6 cells in 1.5 mL of medium in 24-well culture plates: Add Anti-CD3, -CD28 T-Cell expander beads (Dynal/Invitrogen, Carlsbad, USA) at a bead to cell ratio of 1:10 and 100 IU/mL recombinant murine IL-2 (rIL-2). The cells should be restimulated by the addition of fresh rIL-2 every 3 days and split to maintain a concentration of 1×10^6–5×10^6 cells per well. Using a Dynal Particle Concentrator, the Anti-CD3, -CD28 T-Cell expander beads should be removed from culture and replaced with fresh beads every 8 days (*see* **Note 5**).

3.2 Isolation of CD11c+ DC

1. Preceding the mechanical disruption of murine spleens, collagenase D is in injected into the splenic tissue using a sterile syringe (*see* **Note 6**). A single cell suspension of murine splenocytes is then prepared as described under Subheading 3.1 and resuspended in an appropriate volume of MACS buffer.
2. The isolation of CD11c+ splenocytes is a positive selection procedure. The splenocyte single cell suspension is labeled using CD11c MicroBeads (Miltenyi Biotec; Bergisch Gladbach, Germany) according to the manufacturer's protocol.
3. For use in suppression assays, 1×10^4 freshly isolated splenic CD11c+ DC are sufficient for the induction of proliferation of Teff in the presence of soluble anti-CD3 antibody in a single well (*see* Table 1). Alternatively, 2×10^4 T-cell depleted splenocytes and bone marrow derived DC (both FLT3 and GMCSF stimulated, mature DC) have been successfully used as APC fraction in suppression assays [19, 20].

3.3 Suppression Assay

1. Suppression assays are set up in sterile 96-well polystyrene flat bottom cell culture plates. Each plate can harbor 60 individual conditions; the outer rows are not used due to inferior culture conditions.
2. The basic module of a suppression assay includes four conditions, each laid out in a volume of 200 μL and in triplicate for a total of 12 wells (Fig. 1). Cultures are performed in a CO_2 incubator (37 °C, 5 % CO_2, 95 % humidity) for a total of 60 h.
   - Condition 1 includes only 7.5×10^4 CD4+ CD25− Teff and serves as a control of the background proliferation of unstimulated Teff.
   - Condition 2 includes 7.5×10^4 CD4+ CD25− Teff and 1×10^4 DC. It serves as control for Teff proliferation in the presence of unstimulated DC.
   - Condition 3 includes 7.5×10^4 CD4+ CD25− Teff, 1×10^4 CD11c+ DC and a TCR-stimulus. This allows estimation of the proliferation of TCR-stimulated Teff in the presence of DC without Treg.

H	G	F	E	D	C	B	A1
	1. 75.000 CD4+ 2. NO Antibody 3. NO DC 4. NO Treg 5. NO TLR-Ligand	1. 75.000 CD4+ 2. NO Antibody 3. 10.000 DC 4. NO Treg 5. NO TLR-Ligand	1. 75.000 CD4+ 2. αCD3 Antibody 3. 10.000 DC 4. NO Treg 5. NO TLR-Ligand	1. 75.000 CD4+ 2. αCD3 Antibody 3. 10.000 DC 4. 50.000 Treg 5. NO TLR-Ligand			2
	1. 75.000 CD4+ 2. NO Antibody 3. NO DC 4. NO Treg 5. NO TLR-Ligand	1. 75.000 CD4+ 2. NO Antibody 3. 10.000 DC 4. NO Treg 5. NO TLR-Ligand	1. 75.000 CD4+ 2. αCD3 Antibody 3. 10.000 DC 4. NO Treg 5. NO TLR-Ligand	1. 75.000 CD4+ 2. αCD3 Antibody 3. 10.000 DC 4. 50.000 Treg 5. NO TLR-Ligand			3
	1. 75.000 CD4+ 2. NO Antibody 3. NO DC 4. NO Treg 5. NO TLR-Ligand	1. 75.000 CD4+ 2. NO Antibody 3. 10.000 DC 4. NO Treg 5. NO TLR-Ligand	1. 75.000 CD4+ 2. αCD3 Antibody 3. 10.000 DC 4. NO Treg 5. NO TLR-Ligand	1. 75.000 CD4+ 2. αCD3 Antibody 3. 10.000 DC 4. 50.000 Treg 5. NO TLR-Ligand			4

Fig. 1 Schematic illustration of a modular standard assay layout based on a 96-well culture plate. The basic module includes four conditions (column *G–D*), which are repeated once without ligand for the negative control, once with a well-established immunostimulatory ligand (e.g., CpG) for positive control and once for every experimental condition. This modular approach to suppression assays allows a flexible adaption to different experimental settings including the use of TLR ligands, different dendritic cells, viruses, and conditioned medium. Further, standardization of highly repetitive steps reduces the chance of methodological errors

   - Condition 4 includes 7.5×10^4 CD4+ CD25− Teff, 1×10^4 CD11c+ DC, 5.0×10^4 CD4+ CD25+ Treg and a TCR-stimulus. This serves to measure the effect of Treg on the proliferation of the Teff fraction.

3. To induce proliferation of the co-culture, different TCR stimuli may be used (*see* Table 1). Soluble αCD3 antibody (e.g., clone 500A2, BD Biosciences, San Jose, USA) requires binding by dendritic cells to provide a stable interface for T-cell stimulation. In assays where no addition of DC is desired, αCD3 antibody (e.g., clone 145-2C11, BD Biosciences, San Jose, USA) can either be bound directly to the surface of the culture plate or can be covalently bound to commercially available paramagnetic polystyrene MicroBeads (Dynal, Carlsbad, USA; refer to manufacturers protocol for coating instructions).
4. To evaluate the effect of TLR-ligands or other stimuli on the suppressive capacity of Treg, a second set of conditions is set up and the desired ligand is added to each well at the appropriate concentration (*see* **Note 7**).
   - The use of TLR-ligands is well established in the literature (Table 3). A positive control (e.g., LPS or CpG) for the reversal of Treg function should always be included in the assay. Ligands may be directly added to the culture medium

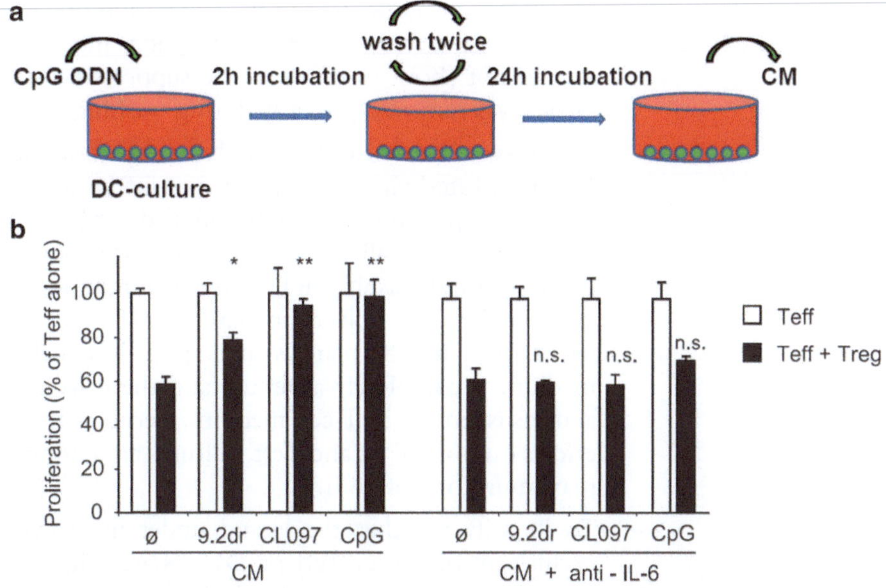

Fig. 2 Schematic illustration of the production of conditioned medium and application in the assay system. (**a**) Conditioned medium was generated by stimulation of 1×10^5 APC with the TLR9 ligand CpG ODN or the TLR7 ligands RNA 9.2dr or CL097 for 2 h. Cells were then spun down and washed twice to remove the ligand, followed by incubation with fresh T-cell medium for 24 h. Supernatants were then carefully harvested and directly transferred into a suppression assay. (**b**) Interestingly, the cell and ligand-free conditioned medium of TLR9 or TLR7 stimulated APC possesses the same potency to abrogate Treg function as the direct stimulation of the coculture with CpG ODN or the synthetic TLR7 ligands 9.2dr RNA or CL097 [11]. This indicates that a soluble factor produced by TLR7 or TLR9-activated APC is sufficient for complete abrogation of Treg-mediated suppression. Suppression can be restored by the addition of neutralizing anti-IL-6 antibody, indicating that IL-6 is an essential factor for the abrogation of Treg mediated suppression by TLR-stimulated APC

and should be carefully resuspended. Immunostimulatory RNA or DNA may also be added directly to the assay using an appropriate transfection medium such as DOTAP™ (Roche Diagnostics; Penzberg, Germany) or Lipofectamine 2000™ (Invitrogen; Carlsbad, USA) (*see* **Note 8**). Table 3 shows well-established ligands for suppression assays in the appropriate concentrations.

- Conditioned medium can be generated by stimulating a dendritic cell culture with the desired TLR-ligand for 2 h (Fig. 2). Cells are then washed twice in fresh T-cell medium and cultured for an additional 24 h to allow secretion of cytokines into the culture medium. Cells are then spun down and the medium is carefully aspirated to prevent cellular contamination, purity can be checked by microscopy or FACS. The medium is then transferred into a suppression assay set up simultaneously. To exclude any effects of leftover ligand, DC that are deficient for the PRR under investigation can be used in the generation of conditioned medium in parallel. The conditioned medium gained should not show an effect in the suppression assay, if all ligand was removed properly during washing.

- Suppression assays can be stimulated with both active and UV-inactivated viruses [11, 18]. For infection with active virus, a suppression assay is set up as described, and spun down to remove all FCS-containing culture medium (*see* **Note 9**). After washing in FCS-free medium, viral particles are added to the culture in pure RPMI 1640 and incubated for 2 h. The plate is then spun down, the culture medium is completely aspirated and the cells are washed and resuspended in T-cell medium. Because of cytopathic effects on both Treg and Teff, cultures are continued for a maximum total of 48 h.
- To identify whether the ligand under investigation acts exclusively on Treg, Teff or DC, assays with mixed cell fractions from different genetic backgrounds may be set up. This has been demonstrated for the TLR7 ligand 9.2dr RNA, where a complete loss of Treg function could be observed when only 1×10^4 DC were added to TLR7-deficient Treg and Teff [11]. When the DC fraction was obtained from TLR7-deficient mice and was cultured in the presence of wild type Treg and Teff, no effect could be observed.

3.4 BrdU Proliferation Assay

1. 5-Bromo-2′-deoxyuridin, a synthetic thymidine analogue, is incorporated into newly synthesized DNA in the S-Phase of the cell cycle [20]. For the use in suppression assays, BrdU (concentration 10 μM) is added to the culture at 48 h after initialization of the assay. Cultures are carefully resuspended and continued for another 12 h (*see* **Note 10**).
2. Detection of intra-nuclear BrdU with peroxidase-conjugated anti-BrdU antibody can be performed following fixation and permeabilization of cells and denaturation of DNA following the manufacturer's protocol (BrdU Cell Proliferation chemoluminescence ELISA kit, Roche Diagnostics, Mannheim, Germany). After binding of the peroxidase labeled antibody, the culture plate is incubated with the provided substrate, leading to light emission which can be detected using a multiwell luminometer (*see* **Note 11**). This will provide readout of the total BrdU incorporation of the culture, but does not allow differentiation of the proliferative activity of different cell types.
3. If detection and quantification of incorporated BrdU is desired on the level of individual cell populations, multi-color flow

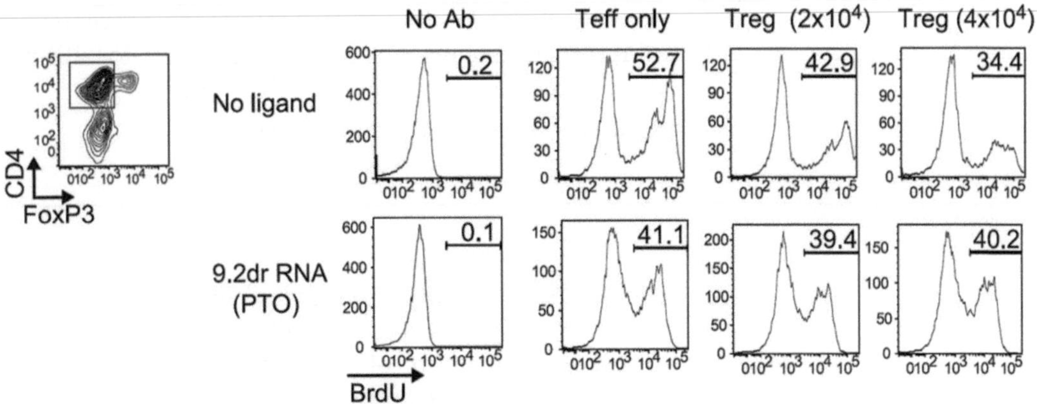

Fig. 3 Visualization of BrdU uptake in individual cell populations of the suppression assay using FACS. To investigate which cell population responds to stimulation of the coculture (Treg, Teff, DC) with the TLR7 ligand RNA9.2dr, the incorporation of BrdU was selectively analyzed in CD4+ FoxP3– Teff using FACS [11]. The results provide conclusive evidence that the proliferation of CD4+ FoxP3– Teff is selectively restored in the presence of TLR7-activated APC

cytometry is preferable (Fig. 3). To this end, a standard surface stain is performed first, according to established protocols. Cells are then fixed and permeabilized using commercially available fixation and permeabilization reagents for FOXP3 staining (FOXP3 Fix/Perm Buffer and FOXP3 Perm Buffer). This allows sequential staining and detection of incorporated BrdU and, if desired, Foxp3 (see below).

- After surface staining, cells are washed in FACS buffer and incubated with 200 μL FOXP3 Fix/Perm Buffer (1:3 concentrate to diluent) for 30 min at 4 °C.
- Spin down the cells and discard the supernatant. Add 200 μL diluted FOXP3 Perm Buffer (1:10 concentrate to distilled H_2O) and resuspend. Spin down cells and discard supernatant. Repeat.
- Add anti-Foxp3 Antibody (*see* Table 3) at a concentration of 1:100 and incubate for 30 min at 4 °C. Spin down cells, discard supernatant and wash twice with 200 μL FOXP3 Perm Buffer.
- Add 200 μL 1 % PFA in PBS and incubate on ice for 10 min at 4 °C. Spin down cells and wash once with 200 μL FOXP3 Perm Buffer.
- Add 200 μL of DNAse I (0.05 mg/mL in PBS; Sigma, Steinheim, Germany) and incubate for 10 min at 37 °C. Continue incubation on ice for an additional 25 min. Spin down cells and wash once with 200 μL FOXP3 Perm Buffer.

- Add anti-BrdU Antibody (ref. Table 3) at a concentration of 1:100 and incubate at room temperature for 20 min. Wash once with 200 μL FOXP3 Perm Buffer.
- Resuspend in FACS Buffer and analyze.

4 Notes

1. Both murine spleens and lymph nodes can be used to isolate functional Treg, Teff and DC. A yield of about 1×10^6 CD4+ CD25+ Tregs at a purity of 95–99 % and about $1–2 \times 10^7$ CD4+ CD25− Teff at a purity of 90–98 % can be expected from a single spleen. Balb/C mice yield tendentially more Treg than the Black/6 strain. A single murine spleen yields about $5 \times 10^5–1 \times 10^6$ CD11c+ DC at a purity of about 80–95 %. Use germ-free mice whenever possible to exclude any effects of prior immune stimulation and infections on Treg function.
2. A test set consisting of three conditions each of Teff and DC without TCR-stimulus, Teff and DC with TCR stimulus and Teff, and DC and Treg with TCR stimulus will require isolation of at least 7.5×10^5 Teff, 1.5×10^5 Treg, and 3×10^4 DC.
3. All steps should be performed without interruption using precooled solutions; all incubations should be performed in the refrigerator at 4 °C. MACS columns need to be prepared by flushing with cold MACS buffer before use.
4. The Treg population is already labeled for CD25 PE, which can be detected by FACS. We strongly recommend checking the purity of the Teff and Treg populations using CD4/CD25 double staining (*see* Table 1 for antibodies) before setting up the suppression assay.
5. Proliferating Treg lose their suppressive function, particularly in the presence of high doses of rIL-2. The expansion of Treg is therefore not useful to gain large numbers of Treg for suppression assays, but rather for PCR analysis or other approaches requiring large numbers of cells.
6. Collagenase digestion increases the yield and purity of the CD11c+ population.
7. The supernatants gained from the assay co-culture can be efficiently used for cytokine-analysis by ELISA or multiplex bead array thereby providing information on the immune micromilieu in the assay and should not be discarded.
8. The transfection rate of primary T-cells is generally low with the use of cationic liposomal transfection reagents. Much rather, the immunostimulatory oligonucleotide added to the assay will primarily be taken up and recognized by dendritic cells. For

improved transfection efficacy of primary T-cells, electroporation may be considered.

9. FCS inhibits the viral infection of responder cells. The infection of the culture conditions therefore has to be performed in FCS-free culture media, e.g., pure RPMI 1640.
10. In standard assay conditions, a total culture time of 60 h with BrdU added after 48 h provides optimum results. When cytotoxic effects of the ligands under investigation on Treg or Teff is a concern, BrdU can be added after a minimum of 24 h of culture, with the readout following 12 h later, for a total culture time of 36 h.
11. Use untransparent culture plates for the suppression assay to reduce background and cross talk of neighboring wells during the substrate reaction.

References

1. Bopp T, Becker C, Klein M et al (2007) Cyclic adenosine monophosphate is a key component of regulatory T cell-mediated suppression. J Exp Med 204:1303–1310
2. Zou W (2006) Regulatory T cells, tumour immunity and immunotherapy. Nat Rev Immunol 6: 295–307
3. Collison LW, Chaturvedi V, Henderson AL et al (2010) IL-35-mediated induction of a potent regulatory T cell population. Nat Immunol 11:1093–1101
4. Hori S, Nomura T, Sakaguchi S (2003) Control of regulatory T cell development by the transcription factor Foxp3. Science 299: 1057–1061
5. Zheng Y, Rudensky AY (2007) Foxp3 in control of the regulatory T cell lineage. Nat Immunol 8:457–462
6. Fontenot JD, Rasmussen JP, Williams LM et al (2005) Regulatory T cell lineage specification by the forkhead transcription factor foxp3. Immunity 22:329–341
7. Thornton AM, Shevach EM (1998) $CD4^+$ $CD25^+$ immunoregulatory T cells suppress polyclonal T cell activation in vitro by inhibiting interleukin 2 production. J Exp Med 188: 287–296
8. Takahashi T, Kuniyasu Y, Toda M et al (1998) Immunologic self-tolerance maintained by $CD25^+CD4^+$ naturally anergic and suppressive T cells: induction of autoimmune disease by breaking their anergic/suppressive state. Int Immunol 10:1969–1980
9. Pasare C, Medzhitov R (2003) Toll pathway-dependent blockade of $CD4^+CD25^+$ T cell mediated suppression by dendritic cells. Science 299:1033–1036
10. Kabelitz D, Wesch D, Oberg HH (2006) Regulation of regulatory T cells: role of dendritic cells and toll-like receptors. Crit Rev Immunol 26:291–306
11. Anz D, Koelzer VH, Moder S et al (2010) Immunostimulatory RNA blocks suppression by regulatory T cells. J Immunol 184:939–946
12. Takeda K, Kaisho T, Akira S (2003) Toll-like receptors. Annu Rev Immunol 21:335–376
13. Libermann TA, Baltimore D (1990) Activation of interleukin-6 gene expression through the NF-k B transcription factor. Mol Cell Biol 10:2327–2334
14. Hornung V, Guenthner-Biller M, Bourquin C et al (2005) Sequence-specific potent induction of IFN-alpha by short interfering RNA in plasmacytoid dendritic cells through TLR7. Nat Med 3:263–270
15. Caramalho I, Lopes-Carvalho T, Ostler D et al (2003) Regulatory T cells selectively express toll-like receptors and are activated by lipopolysaccharide. J Exp Med 197:403–411
16. Peng G, Guo Z, Kiniwa Y et al (2005) Toll-like receptor 8-mediated reversal of $CD4^+$ regulatory T cell function. Science 309: 1380–1384
17. Sutmuller RP, den Brok MH, Kramer M et al (2006) Toll-like receptor 2 controls expansion and function of regulatory T cells. J Clin Invest 116:485–494
18. Kurooka M, Kaneda Y (2007) Inactivated Sendai virus particles eradicate tumors by inducing

immune responses through blocking regulatory T cells. Cancer Res 67: 227–236

19. Wurzenberger C, Koelzer VH, Schreiber S et al (2009) Short-term activation induces multifunctional dendritic cells that generate potent antitumor T-cell responses in vivo. Cancer Immunol Immunother 58:901–913

20. Rußmann E (1993) Colloquium Roche Mol Biochem 4:1–4

21. Crellin NK, Garcia RV, Hadisfar O et al (2005) Human CD4+ T cells express TLR5 and its ligand flagellin enhances the suppressive capacity and expression of FOXP3 in CD4+CD25+ T regulatory cells. J Immunol 175:8051–8059

Chapter 12

MicroRNA Methodology: Advances in miRNA Technologies

Theresa Kaeuferle, Sabine Bartel, Stefan Dehmel, and Susanne Krauss-Etschmann

Abstract

There is an emerging trend in microRNA research and thus substantial progress in microRNA technologies. In this chapter we provide insights into the main microRNA specific methodologies and critical steps of microRNA expression profiling, target gene identification, and functional confirmation of microRNA effects up to in vivo application of antagomirs.

Key words MiRNA overexpression, Luciferase binding assays, In silico target prediction, MiRNA quantification

1 Introduction

During the last years, a large number of non-protein-coding RNAs (ncRNAs) have been identified as fundamental regulatory molecules [1]. MicroRNAs (miRNAs) are one class of small (18–25 nucleotides) ncRNAs that regulate gene expression posttranscriptionally through partial Watson–Crick base-pairing with their target mRNAs, leading to mRNA degradation or repression of translation [2, 3]. MiRNAs have been reported to play regulatory roles in the development and pathogenesis of numerous mammalian organs, like lung, heart, kidney, brain or liver [4]. The accumulating knowledge on these RNAs is leading to the development of more efficient and reliable methods for studying miRNA function. Nevertheless, compared to other nucleic acids miRNA methodology is more challenging due to several factors including the small size of miRNAs, the difficulty to discriminate the different processing states of pre-, pri-, and mature miRNAs, the variable Tm of primers and probes, a high degree of homology in miRNA families, the ongoing discovery of new miRNAs and the inaccuracy of target predictions due to the imperfect pairing of miRNAs to their target mRNAs.

Hans-Joachim Anders and Adriana Migliorini (eds.), *Innate DNA and RNA Recognition*, Methods in Molecular Biology, vol. 1169, DOI 10.1007/978-1-4939-0882-0_12, © Springer Science+Business Media New York 2014

Several miRNA specific methods from identification of candidate miRNAs up to influencing biological processes by their modulation have been developed. Firstly, for high-throughput screening of miRNAs and generation of differential miRNA profiles in a particular experimental setting, miRNA microarrays are used. Performing different microarray technologies helps to narrow down the number of false positive candidate miRNAs. MiRNA microarray results are usually further validated by qRT-PCR of single miRNAs. MiRNA target genes may then be identified by using computational methods. As miRNA binding to their target sequence, which is mostly located in the 3'UTR of target genes, does not require a perfect match, in silico research tools rely on mathematical algorithms to determine high binding probability. For high-throughput experimental identification of putative target genes mRNA array data can be inversely correlated to the miRNA expression data followed by in silico pathway analysis. However, to experimentally confirm miRNA-mediated gene regulation, it is feasible to use luciferase-based binding assays as well as miRNA overexpression or inhibition experiments. Overexpression/inhibition of endogenous miRNA levels in vitro may also give insights into miRNA function in biological processes or may identify new target genes by for example downstream proteome analysis. In the end, miRNA levels can also be modulated in in vivo settings to study their relevance for physiological processes or disease pathogenesis.

2 Materials

2.1 MiRNA Expression Profiling

Prepare all solutions with nuclease-free water, use UV light and surface decontaminating agents to remove nucleases and nucleic acids before starting. Work with RNAse-free material to prevent contamination and RNA degradation.

1. RNAlater® (Ambion, Life Technologies, Carlsbad, CA, USA), Trizol.
2. Isolation kit for total RNA including miRNAs.
3. Ammonium acetate, glycogen, 100 and 70 % ethanol.
4. Capillary gel electrophoresis system to control RNA quantity and quality.
5. Two different microRNA microarray technologies are described here (*see* also **Notes 1** and **2**).

   For miRCURY™ LNA microRNA Arrays: miRCURY™LNA microRNA Array Kit and miRCURY LNA microRNA Array Power Labeling Kit (Exiqon,Vedbaek, Denmark), DMSO, Microarray Hybridization Chamber—SureHyb, Hybridization Gasket Slide Kit, glass staining jar/dish or equivalent, GenePix 4000A Microarray Scanner, and GenePix® Pro Software (Axon Instruments, Foster City, CA, USA).

For TaqMan® Low-Density Arrays: TaqMan® microRNA Array A and B (Applied Biosystems, Megaplex™ RT primers, TaqMan® MicroRNA Reverse Transcription Kit, Megaplex™ PreAmp Primers, TaqMan® PreAmp Master Mix, TaqMan® Universal PCR Master Mix (No AmpErase® UNG), ABI PRISM® 7900HT Fast Real-Time PCR System, SDS software (all from Applied Biosystems, Life Technologies, Carlsbad, CA, USA).

6. *For TaqMan® qRT-PCR*: TaqMan® MicroRNA Assays containing 5× TaqMan® MicroRNA Assay RT Primer and 20× TaqMan® MicroRNA Assay PCR Primer; *as for TaqMan® Low-Density Arrays:* TaqMan® MicroRNA Reverse Transcription Kit, TaqMan® Universal PCR Master Mix (No AmpErase® UNG), ABI PRISM® 7900HT Fast Real-Time PCR System, SDS software (all from Applied Biosystems).

2.2 Luciferase Reporter Assays

1. Appropriate cell line (*see* **Note 3**).
2. PCR purification kit.
3. Bi-cistronic reporter plasmid (*see* **Note 4**).
4. miRNA molecules (*see* **Note 5**).
5. Cell culture medium: Growth medium depending on the cell type.
6. Transfection medium: serum-free growth medium for lipid-based transfection to allow generation of lipid complexes.
7. Lipofectamine 2000 (Invitrogen, Life Technologies, Carlsbad, CA, USA).
8. 24-Well cell culture plates.
9. Microplate for luminescence based assays.
10. Lysis buffer: 250 mM Tris, 1 % Triton, pH 7.7. Weigh 15.14 g Tris, add 400 ml water and adjust pH with HCl. Make up to 500 ml with water. Add 5 ml Triton (100×). Store at RT. For working solution dilute 1:5 with water.
11. Substrates for luciferases, exemplary for *Renilla* (coelenterazine) and firefly (luciferin) luciferases:

    Coelenterazine: Dissolve 10 mg coelenterazine in 1 ml MeOH. Take 69.32 μl and add 32.93 ml PBS. Store at −20 °C and protect from light. For working solution dilute 1:5 with PBS.

    Luciferin: Weigh 100 mg D-Luciferin, 159.1 mg Coenzyme A, 3,894 mg DTT, 221 mg ATP, 349 mg $(MgCO_3)_4Mg(OH)_2 \cdot 5H_2O$, 498 mg $MgSO_4 \cdot 7H_2O$, 2.715 g Tricine, 28.2 mg EDTA. Make up to 757 ml with water. Perform 30 min sonification. Store at −20 °C and protect from light.

2.3 MiRNA Overexpression and Knockdown

1. Appropriate cell line (*see* **Note 3**).
2. miRNA mimics/inhibitors (*see* **Note 5**).
3. Cell culture medium: Growth medium depending on the cell type.
4. Transfection medium: serum-free growth medium for lipid-based transfection to allow generation of lipid complexes.
5. PBS.
6. 24-Well cell culture plates.

3 Methods

3.1 MiRNA Expression Profiling

1. Prepare samples for isolation as followed:

   Tissue samples: Store tissue samples in RNAlater® until further processing to prevent degradation. Transfer tissue to Trizol reagent and homogenize.

   Blood samples: Transfer whole blood directly to Trizol solution (approximate ratio 1:6).

   Cell culture: Dissolve cell culture samples by directly adding Trizol.
2. For isolation of total RNA use an isolation kit that includes miRNAs. For miRNA microarrays further purify total RNA by precipitation adding 10 μl ammonium acetate, 1 μl glycogen, and 450 μl 100 % ethanol per 100 μl RNA solution, followed by incubation at -20 °C overnight. The precipitate is then centrifuged, washed two times with 70 % ethanol, and resuspended in nuclease-free water.
3. Monitor RNA quality and quantity (e.g., by using the Bioanalyzer system). RNA integrity numbers (RIN) above 7 indicate sufficient RNA quality.
4. For *miRCURY™LNA microRNA* Arrays hybridize Hy3-/Hy5 labeled RNA manually on miRNA microarrays according to manufacturer's instructions.
   - Combine 1 μg total RNA with spike-in miRNAs, CIP buffer, and CIP enzyme and incubate for 30 min at 37 °C followed by heating for 5 min at 95 °C. Cool on ice for 2 min.
   - Attach fluorescent labels (Hy3™ or Hy5™, respectively) using labeling enzyme, labeling buffer, and DMSO for 1 h at 16 °C. Stop reaction by 15 min at 65 °C. Mix the two samples and denature with hybridization buffer for 2 min at 95 °C followed by 2 min on ice.
   - Pipet RNA preparation into prepared microarray slide and incubate for hybridization within a tightly closed slide chamber for 16–18 h at 56 °C.

- Wash for 2 min in buffer A at 56 °C, 2 min in buffer B at 23 °C and 2 min in buffer C at 23 °C. Subsequently dry slides.
- Measure fluorophore emissions of 556 nm (Hy3™) and 656 nm (Hy5™) by scanning the slides at 5 µm by a GenePix Microarray Scanner. Match fluorescent signals to species-specific GenePix® Array List (GAL) files.
- Analyze by GenePix® Pro Software. Remove non-positive spots after background correction. Normalize the remaining signal intensity values to per-chip median values. Test repeatability between replicate spots with color swap.

5. Perform *TaqMan® Low-Density Arrays* according to the supplier's protocol, including pre-amplification steps.
   - Subject 350 ng total RNA to RT reaction including Megaplex™ RT primers and TaqMan® MicroRNA Reverse Transcription Kit. Set cycling conditions: 40 cycles 2 min at 16 °C, 1 min at 42 °C, 1 s at 50 °C; final heating 5 min at 85 °C; cool down to 4 °C.
   - Combine with Megaplex™ PreAmp Primers and TaqMan® PreAmp Master Mix: 10 min at 95 °C, 2 min at 55 °C, 2 min at 72 °C; 12 cycles: 15 s at 95 °C, 4 min at 60 °C; 10 min at 99 °C; cool down to 4 °C.
   - Mix diluted pre-amplified RT product with TaqMan® Universal PCR Master Mix. Load on 384-well microfluidic card of the TaqMan® Array, centrifuge and seal. Perform TaqMan® LDA runs on an ABI PRISM® 7900HT Fast Real-Time PCR System.
   - Analyze by SDS software employing the $\Delta\Delta C_T$-method. Choose an appropriate reference gene for normalization or use global mean normalization [5].
6. Confirm differential expression of single candidate miRNAs using *qRT-PCR* (*see* **Note 6**).
   - Perform multiplex RT reaction with miRNA specific TaqMan® MicroRNA RT primer pool and TaqMan® MicroRNA Reverse Transcription Kit according to the manufacturer's guidelines. Therefore, combine 10 µl of each individual 5× TaqMan® MicroRNA Assay RT primer and add 1× TE buffer to bring the final volume to 1,000 µl. Mix 10× Reverse Transcription Buffer, RT Primer pool, dNTPs, MultiScribe™ Reverse Transcriptase, RNase inhibitor, and 350–1,000 ng total RNA in a 15 µl reaction volume. Incubate on ice for 5 min. Set cycling conditions: 30 min at 16 °C, 30 min at 42 °C, 5 min at 85 °C. Store at −20 °C.
   - Quantify miRNA levels in a real-time PCR reaction using 20× TaqMan® MicroRNA Assay PCR Primer, 2× TaqMan®

Universal PCR Master Mix, and 0.16 μl of cDNA in a 20 μl reaction volume. Set cycling conditions: 10 min at 95 °C, 40 cycles: 15 s at 95 °C, 60 s at 60 °C.

- Choose an appropriate reference gene for normalization (*see* **Note 7**) and analyze by $\Delta\Delta C_T$-method.

3.2 *In Silico* Target Prediction

1. Binding site predictions often lead to high numbers of putative target genes for a single miRNA. To avoid too many false positive target genes, use a strict "consensus approach" based on different prediction algorithms. Since miRNAs from the same family share their seed region, you may group them for target prediction. Perform target predictions using following algorithms as an example (*see* **Note 8**):

   It might be useful to focus on targets with a consensus target prediction of at least 3 different algorithms. Use the online database miRWalk (http://www.umm.uni-heidelberg.de/apps/zmf/mirwalk/), which also offers information on predicted and validated targets for combined use of target prediction algorithms [6].

2. For enrichment analysis of target genes in specific pathways use DIANA-mirPath software (http://diana.cslab.ece.ntua.gr/pathways/) [7].

3.3 Luciferase Reporter Assays

Carry out all procedures at room temperature under sterile conditions in a cell culture hood unless otherwise specified. Include a vector-only control as well as co-transfection of vector and negative control miRNA. You may also use plasmids with mutated miRNA binding sites for control.

1. Amplify the 3'UTR of the target gene by PCR and verify the product via gel electrophoresis (*see* **Note 9**).
2. Purify PCR product.
3. Clone the amplified 3'UTR into the multiple cloning site of a bi-cistronic reporter vector and verify construct by control digestion or sequencing (*cloning is not described here*).
4. Seed cells in a 24-well multidish the day before transfection, so that a confluency of at least 30 % is reached at time of transfection. Use 500 μl of cell culture medium per well and incubate cells at standard conditions (37 °C and 5 % CO_2, or depending on cell type) (*see* **Note 10**).
5. Co-transfect 20 pmol of the respective miRNA or negative control and 100 ng vector construct per well (24-well plate) using Lipofectamine 2000 (*see* **Note 11**; *for transfection see* Subheading 3.4).

6. 72 h post transfection (*see* **Note 12**) lyse cells by addition of 120 μl of lysis buffer per well and incubate 10 min at room temperature with constant agitation. For luciferase measurement pipet each 50 μl in appropriate 96-well plates. Measure reporter and transfection-control luciferase activities after addition of 100 μl substrate per well.
7. For each sample, normalize reporter luciferase activity to transfection-control luciferase activity. Positive binding of miRNAs to the gene will result in lowered reporter/ transfection-control luciferase activity ratios.

3.4 MiRNA Overexpression/ Knockdown (Transient)

In the following section transfection with lipid-based technology using Lipofectamine 2000 reagent is described (*see* **Note 13**).

1. Seed cells as for luciferase reporter assays (*see* Subheading 3.3).
2. Prepare a miRNA mix containing 20 pmol miRNA mimic/ inhibitor (*see* **Note 11**) in 50 μl transfection medium per well and mix gently.
3. Prepare a lipofectamine mix adding 2 μl of Lipofectamine 2000 to 48 μl of transfection medium per well. Mix gently and incubate at room temperature for 5 min.
4. Mix miRNA mix with Lipofectamine mix and incubate at room temperature for 20 min. Prepare a miRNA–Lipofectamine "mastermix" for replicates. Do not vortex!
5. Aspirate culture medium from cells and wash cells with PBS.
6. Add 100 μl of the miRNA–Lipofectamine mix and 400 μl growth medium to each well. Culture the cells for 24 h (inhibition assays) or 72 h (overexpression and luciferase assays) until analysis (*see* **Note 12**).
7. Validate the efficiency of miRNA overexpression/inhibition by qRT-PCR (*see* Subheading 3.1). Proceed to regular RNA or protein isolation techniques from cell culture samples for evaluating endogenous target transcript/protein levels by mRNA qRT-PCR or western Blot.

3.5 In Vivo Application

MiRNA levels can be inhibited in vivo for example by application of "antagomirs" which are single stranded, chemically modified, and cholesterol conjugated antisense molecules [8]. They can be administered intravenously [8] or for example intranasally [9, 10] or intratracheally [11] to explore their function in lung diseases. Additionally, the locked nucleic acid (LNA) technology among others has shown successful inhibition of endogenous miRNA levels (reviewed in [12]). Most of the modified antisense molecules are already commercially available.

4 Notes

1. For initial profiling we recommend to use two miRNA microarrays based on different technologies. During the last years two technologies have emerged: hybridization-based (e.g., Affymetrix, Agilent, Ambion, Exiqon, Invitrogen, and Toray) and PCR based (e.g., Exiqon, TaqMan®) microarrays. In this chapter miRCURY™LNA microRNA Arrays and TaqMan® Low-Density Arrays are exemplary described. There are a few studies [13–15] comparing different commercially available miRNA microarrays concluding that intra-platform reproducibility is relatively stable whereas inter-platform reproducibility among different platforms is low. You may use the help of these studies for selection of the platform which is most suitable for you. Note that different commercially available microarrays include different sets of miRNAs that are measured.
2. Recent introduction of deep sequencing technology has provided another option for miRNA profiling overcoming microarray disadvantages like background and cross-hybridization problems and the measurement of only previously discovered miRNAs. Thus deep sequencing allows the discovery of novel miRNAs and other small RNA species and measures absolute abundance. Disadvantages are the need of professional bioinformatics support, cost, labor, and time consumption.
3. The cell line you choose depends on the aim of your experiment: For simple interaction studies between miRNA and putative target genes you can do your experiments with easy-to-handle cells like HEK-293. For identifying biological function you may choose tissue-specific cells or even primary cells.
4. Bi-cistronic reporter vectors contain the genomic sequence of two luciferase genes. Expression of one luciferase is directly linked to the cloned 3'-UTR of your gene of interest to evaluate miRNA regulation, whereas an independently expressed second luciferase can be used as transfection control. Positive binding of miRNAs to the gene will result in lowered luciferase signal ratios. This system keeps you from extensive experiments for protein content measurements to normalize for cell density and allows you to correct your results for transfection efficiencies.
5. MiRNA molecules, like miRNA mimics or inhibitors are already commercially available for a whole lot of specific miRNAs. Some companies offer a custom-made service if there is no predesigned molecule for your miRNA of interest. We use mirVana™ miRNA-mimics (Life Technologies) or Pre-miR™ miRNA Precursors (Ambion) and mirVana™ miRNA inhibitors (Life Technologies) with the appropriate commercially available negative controls.

6. qRT-PCR has been widely used as "gold standard" for miRNA quantification and thus is a useful tool to validate miRNA microarray results. There are different types of miRNA qRT-PCRs, such as stem-loop RT-primer (TaqMan®), locked nucleic acid primers (Exiqon) and qRT-PCR with poly-A tailing (QIAGEN, Stratagene). TaqMan® microRNA assay system is described here, since it seems the most widely used. To overcome the disadvantage of miRNA specific RTs, TaqMan newly offers a protocol for a multiplex RT step to generate PCR quantification out of one cDNA pool. With this protocol up to 96 primers can be pooled into one RT reaction.
7. Different reference genes for miRNA analysis are commercially available. Mostly these are rather small nuclear RNAs like for example sno-234 (murine) or RNU-48 (human). As these are not miRNAs per se an alternative approach is to use a combination of 3–5 non-regulated miRNAs from your previously performed microarray which resemble the mean expression level [16].
8. There are first-generation and second-generation methods of prediction algorithms classified according to their characteristic properties. For more details see [17].
9. If the 3'UTR of your target mRNA is very long, use a polymerase with proofreading function to avoid mutations in the fragment and prolong extension time to 2.6 min per kb. As very big plasmids might be difficult to transform, you may just clone the miRNA binding site region into the reporter vector instead of complete 3'UTR.
10. Cell density and passage number of the cells can be crucial for the outcome of your experiment.
11. It might be useful to vary miRNA concentrations (or plasmid/miRNA ratios in luciferase reporter assays) and the amount of Lipofectamine 2000 per well to achieve higher transfection efficiencies. Perform a dosage finding experiment at first.
12. Downregulation of luciferase signal or transcript levels in miR-transfected cells is time-dependent. Though 48 h/72 h seem standard you may have to test longer or shorter time-points in pre-experiments. For culture longer than 72 h change medium after 48 h.
13. Some cell lines and especially primary cells may require different transfection techniques like electroporation, nucleofection, or magnetofection. Immortal cell lines are usually easy to transfect with lipid-based technologies (We use Lipofectamine® 2000 or lipofectamine® RNAiMAX (Life technologies, Darmstadt, Germany)). Besides, there are an increasing number of commercially available transfection reagents, like ExGen (Fermentas, Thermo Fisher Scientific), FuGENE® HD

Transfection Reagent (Promega), LipoGen™ (InvivoGen, San Diego, CA, USA), Effectene Transfection Reagent, SuperFect® Transfection Reagent, and Polyfect® (all from Qiagen, Hilden, Germany), and ESCORT™ Transfection Reagent (Sigma-Aldrich, St. Louis, MO, USA).

References

1. Amaral PP, Dinger ME, Mercer TR et al (2008) The eukaryotic genome as an RNA machine. Science 319:1787–1789
2. Huntzinger E, Izaurralde E (2011) Gene silencing by microRNAs: contributions of translational repression and mRNA decay. Nat Rev Genet 12:99–110
3. Jackson RJ, Standart N (2007) How do microRNAs regulate gene expression? Sci STKE 2007:re1.
4. Sayed D, Abdellatif M (2011) MicroRNAs in development and disease. Physiol Rev 91: 827–887
5. D'Haene B, Mestdagh P, Hellemans J et al (2012) miRNA expression profiling: from reference genes to global mean normalization. Methods Mol Biol 822:261–272
6. Dweep H, Sticht C, Pandey P et al (2011) miRWalk-database: prediction of possible miRNA binding sites by "walking" the genes of three genomes. J Biomed Inform 44: 839–847
7. Papadopoulos GL, Alexiou P, Maragkakis M et al (2009) DIANA-mirPath: integrating human and mouse microRNAs in pathways. Bioinformatics 25:1991–1993
8. Krutzfeldt J, Rajewsky N, Braich R et al (2005) Silencing of microRNAs in vivo with 'antagomirs'. Nature 438:685–689
9. Mattes J, Collison A, Plank M et al (2009) Antagonism of microRNA-126 suppresses the effector function of TH2 cells and the development of allergic airways disease. Proc Natl Acad Sci U S A 106:18704–18709
10. Collison A, Mattes J, Plank M et al (2011) Inhibition of house dust mite-induced allergic airways disease by antagonism of micro RNA-145 is comparable to glucocorticoid treatment. J Allergy Clin Immunol 128:160.e4–167.e4
11. Pandit KV, Corcoran D, Yousef H et al (2010) Inhibition and role of let-7d in idiopathic pulmonary fibrosis. Am J Respir Crit Care Med 182:220–229
12. Stenvang J, Kauppinen S (2008) MicroRNAs as targets for antisense-based therapeutics. Expert Opin Biol Ther 8:59–81
13. Wang B, Howel P, Bruheim S et al (2011) Systematic evaluation of three microRNA profiling platforms: microarray, beads array, and quantitative real-time PCR array. PLoS One 6:e17167
14. Git A, Dvinge H, Salmon-Divon M et al (2010) Systematic comparison of microarray profiling, real-time PCR, and next-generation sequencing technologies for measuring differential microRNA expression. RNA 16: 991–1006
15. Sato F, Tsuchiya S, Terasawa K et al (2009) Intra-platform repeatability and inter-platform comparability of microRNA microarray technology. PLoS One 4:e5540
16. Mestdagh P, Van Vlierberghe P, De Weer A et al (2009) A novel and universal method for microRNA RT-qPCR data normalization. Genome Biol 10:R64
17. Li L, Xu J, Yang D et al (2010) Computational approaches for microRNA studies: a review. Mamm Genome 21:1–12

Part II

Analysis of Nucleic Acid Sensing In-Vivo

Chapter 13

Expression Profiling by Real-Time Quantitative Polymerase Chain Reaction (RT-qPCR)

Maciej Lech and Hans-Joachim Anders

Abstract

Real-time quantitative PCR is a variation of the standard PCR technique that is commonly used to quantify nucleic acid. However, in this technique the amount of amplified specific sequence can be quantified at each stage of the PCR cycle. If investigated sequence is present in large number of copies in particular sample, amplification product is detected already in earlier cycles; if the sequence is rare, amplification is observed in later cycles. Quantification of amplified product is acquired using fluorescent probes or fluorescent DNA-binding dyes. Accumulation of fluorescent signal can be measured by real-time PCR instruments during each of 35–45 cycwwles of the PCR reaction, which simplify the procedure by eliminating the visualization of the amplified products using gel electrophoresis. Real-time-PCR allows quantifying the amount of product already during the PCR reaction as soon as it is detectable. Correctly performed, this method may be used for precise gene expression analysis in life science, medicine, and diagnostics and has become the standard method of choice for the quantification of mRNA. However in the past few years it became obvious that real-time PCR is complex and variability of RNA templates, assay designs, inappropriate data normalization, and data interpretation may cause diverse analytical problems.

Key words RT-PCR, Quantification, Polymerase chain reaction, Primer design

1 Introduction

Quantitative nucleic acid sequence analysis plays an important role in biological research. Measurement of gene expression (RNA) is frequently used for monitoring biological responses to various stimuli as well as to demonstrate changes in biological responses throughout the different phases of the disease. Among many methods used for the investigation of nucleic acids abundance, PCR has been proven to be a powerful tool for quantification. Older detection systems such as agarose gels, fluorescent labeling of PCR products, or probe hybridization require post-PCR manipulation and are not only less precise, time-consuming, and often require

Hans-Joachim Anders and Adriana Migliorini (eds.), *Innate DNA and RNA Recognition*, Methods in Molecular Biology, vol. 1169, DOI 10.1007/978-1-4939-0882-0_13, © Springer Science+Business Media New York 2014

larger amount of samples [1–3]. The first practical kinetic PCR assay was established in 1993 and combined amplification of nucleic acid with monitoring of newly synthesized DNA. Assay was based on 5′ nuclease activity of *Taq* polymerase to cleave a nonextendible hybridization probe during the extension phase of PCR and required dual-labeled fluorogenic hybridization probes: FAM (6-carboxyfluorescein) and TAMRA (6-carboxy-tetramethylrhodamine) [4]. The principle of this technique did not change over the last years. Quantitative PCR assay depends on measuring the increase in fluorescent signal, which is proportional to the amount of DNA synthesized during each PCR cycle. However present tests allow quantification of multiple target genes in a single reaction (probes labeled with different dyes). Individual reactions are analyzed by the PCR cycle at which fluorescence rises above background fluorescence (threshold), a parameter known as the threshold cycle (Ct) or crossing point (Cp). Samples which contain higher amount of the target give lower Ct/Cp values. The correlation between fluorescence and amount of amplified product allows quantification of target. In last few years one of the most common quantitative measurements of ds-DNA in solution is achieved using SYBR Green I assay system for detection. This system allows the specific detection of ds-DNA over a broad concentration range, using a single low dye concentration. Moreover SYBR Green I fluorescence is not significantly affected by pH variation and reliable in samples contaminated with typical salts and other substances which may be present due to the RNA/cDNA preparation process. During the extension phase SYBR green binds to the PCR product, resulting in increased fluorescence [5]. Consequently during each PCR cycle more fluorescence signal will be detected. In the initial PCR cycles the fluorescent signal is too weak to be detected and it is usually registered after around 10 cycles. Fluorescence doubles at each cycle and began to plateau as PCR process reaches the saturation status. Ct/Cp value is proportional to the logarithm of initial amount of the target in a sample. The differences in the cycle number after which same target in the treated sample was detected indicate the difference in expression level. Only correctly designed assays and optimal designed primers allow generating reliable and reproducible data. Features as primer-dimers formation, lack of specificity of primers, and formation of secondary structures in the PCR products might lead to increase of background fluorescence. Standardization of every step is essential to generate quantitative data that are reliable and reproducible. Moreover RT-qPCR data gives only information about the quantity of a particular transcript in a cell or tissue. Issues such as post-transcriptional regulation, microRNA, protein level, and activity must be considered.

2 Materials

2.1. Chemicals and laboratory equipment: sterile.

1. RNase-free pipette tips.
2. RNA-later solution.
3. 2-Merkaptoethanol.
4. 70 % ethanol (in RNAse-free water).
5. 96–100 % ethanol.
6. RNase-free molecular biology grade water.
7. Agarose.
8. Sodium acetate.
9. 8-Methoxypsoralen (MOPS).
10. EDTA.
11. 37 % formaldehyde.
12. Ethidium bromide.
13. Bromophenol blue.
14. 100 % glycerol.
15. Formamide.

2.2. Total RNA purification kit.

2.3. DNase on-column digestion kit.

2.4. 2× SYBR green mix.

2.5. 10 μM reverse and forward primers.

2.6. Reverse transcription reaction.

1. Reverse Transcriptase.
2. 5× first-strand buffer.
3. 25 mM deoxyribonucleotide triphosphates (dNTPs).
4. 0.1 M dithiothreitol (DTT).
5. Random primers.
6. Thermo stable RNase inhibitor.

2.7. Gel and running buffer components.

1. 5× loading buffer (16 μl saturated aqueous bromophenol blue solution, 80 μl 0.5 M EDTA, pH 8.0, 720 μl 37 % (12.3 M) formaldehyde, 2 ml 100 % glycerol, 3,084 μl formamide, 4 ml 10× formaldehyde agarose gel buffer, RNase-free water to 10 ml).
2. Formaldehyde agarose gel running buffer (100 ml 10× formaldehyde agarose gel buffer, 20 ml 37 % (12.3 M) formaldehyde, 880 ml RNase-free water).

2.8. Master mix for cDNA synthesis (values for one reaction).

1. 4 µl Reverse Transcriptase.
2. 5× first-strand buffer.
3. 0.4 µl 25 mM dNTPs.
4. 1 µl 0.1 M DTT.
5. 0.215 µl random primers.
6. 0.5 µl thermo stable RNAse inhibitor (40 units/µl).
7. 0.435 µl Reverse Transcriptase.
8. 3.45 µl RNAse-free water.

3 Methods

3.1 RNA Extraction (Sample Size Normalization)

1. Store the harvested tissue immediately following sacrifice (*see* **Note 1**).
2. Use the RNAse-free tubes and spatula while preparing samples for homogenization. This step should be used as a first step for the normalization. It is recommended to use same amount of cells and organs for the RNA extraction (up to 200 mg of animal tissue) and homogenize it with handheld, rotor-stator homogenizer in appropriate volume of lysis buffer containing 1 % 2-mercaptoethanol (350–500 µl) for 20–40 s on ice. Centrifuge sample at 2,000–3,000 ×*g* for 5 min at room temperature; pipette the supernatant into new tube. Add same volume of 70 % RNAse-free ethanol to the tissue homogenate and vortex it few seconds.
3. Transfer sample to the Spin Cartridge and centrifuge at 12,000 ×*g* for 15 s.
4. Use on-column DNase treatment during the purification procedure.
5. Add 700 µl wash buffer I to the Spin Cartridge; centrifuge at 12,000 ×*g* for 15 s.
6. Add 500 µl wash buffer II to the Spin Cartridge; centrifuge at 12,000 ×*g* for 15 s (repeat this step).
7. Elute the RNA from the Spin Cartridge using RNAse-free water (30–100 µl).
8. Store your purified RNA on ice. For long-term storage, store the purified RNA at −80 °C.

3.2 RNA Quality Check

The most common method used to assess the integrity of total RNA is agarose gel stained with ethidium bromide. Intact total eukaryotic RNA run on agarose gel will have sharp, clear 28S (~4,800 bases) and little less intense 18S (~1,900 bases) rRNA bands (2:1 ratio).

Degraded RNA will have a smeared appearance, will lack the sharp rRNA bands, and exhibit untypical 28S:18S ratio or run at low molecular weight.

1. Prepare Formaldehyde 1.2 % Agarose gel in formaldehyde buffer (20 mM MOPS, 5 mM sodium acetate, 1 mM EDTA, pH to 7.0 with NaOH).
2. Heat the mixture to melt agarose, cool to 60 °C water bath, and add 1.8 ml of 37 % (12.3 M) formaldehyde and 1 μl of a 10 mg/ml ethidium bromide. Mix thoroughly and pour onto gel support (*see* **Note 2**).
3. Prepare the RNA samples by adding 1 volume of 5× loading buffer per 4 volumes of RNA sample. Incubate for 5 min at 65 °C, chill on ice, and load onto the formaldehyde agarose gel.
4. Run gel at 5–7 V/cm in 1× formaldehyde agarose gel running buffer until the leading bromophenol blue dye front has migrated approximately three quarters of the length of the gel. Visualize RNA by irradiation with UV light (*see* **Note 3**).

3.3 Reverse Transcription, cDNA Synthesis (Second Normalization Step: Equal Amount of RNA)

1. Prepare the master mix, accounting for one or two additional reactions per gene.
2. Prepare 1 μg of total RNA in 10 μl RNase-free water, heat it to 65 °C for 5 min and chill on ice. Same amount of good quality RNA should be preceded to cDNA.
3. Centrifuge RNA briefly and mix it gently with 10 μl Master Mix for cDNA synthesis.
4. Incubate at 25 °C for 10 min (to allow the random hexamers to anneal).
5. Incubate at 42 °C for 90 min.
6. Inactivate the reaction by heating at 70 °C for 15 min.

3.4 Real-Time Quantitative PCR

The synthetized cDNA can now be used as a template for amplification in PCR. Each PCR reaction (usually 20 μl) contains: 10 μl 2× SYBR Green reagent with Taq polymerase, 0.6 μl forward primer (10 μM stock), 0.6 μl reverse primer (10 μM stock), 0.2 μl cDNA, and 8.6 μl Molecular biology grade water. SYBR Green Dye detection system can be used for amplification (*see* **Note 4**). Each amplification step included initiation phase 95 °C, annealing phase 60 °C, and amplification phase 72 °C and was repeated 45 times. Controls consisting of ddH2O (no template controls) and samples synthetized without reverse transcriptase (non-RT reactions should be prepared according to **steps 1–6** but without enzyme) should be included to rule out the unspecific or background signal. No Template Controls (NTCs) should be included on every plate for every primer set to check for contamination.

	HKG 1	HKG 2	HKG 3	HKG 4	HKG 5	HKG 6
A 1	**20,17**	17,48	**18,98**	21,84	15,78	**7,66**
A 2	**20,54**	16,85	**19,78**	23,84	16,78	**8,05**
A 3	**20,75**	18,95	**21,03**	22,84	15,28	**9,22**
A 4	**19,31**	18,44	**20,08**	19,75	19,85	**7,69**
A 5	**21,03**	17,59	**20,58**	20,87	20,78	**8,39**
A 6	**20,21**	18,63	**19,76**	22,08	20,96	**8,26**
B 1	**19,65**	19,45	**19,57**	19,75	19,24	**8,8**
B 2	**19,57**	16,99	**19,78**	19,07	21,07	**7,61**
B 3	**20,47**	19,41	**20,74**	19,53	15,45	**8,79**
B 4	**20,75**	20,01	**20,28**	18,75	16,44	**7,33**
B 5	**21,38**	18,77	**20,97**	21,45	19,54	**7,56**
B 6	**20,22**	19,25	**19,87**	20,78	19,21	**8,42**
N	**11**	11	**11**	11	11	**11**
GM	**20,33**	18,46	**20,11**	20,83	18,24	**8,13**
AM	**20,34**	18,49	**20,12**	20,88	18,37	**8,15**
Min.	**19,31**	16,85	**18,98**	18,75	15,28	**7,33**
Max.	**21,38**	20,01	**21,03**	23,84	21,07	**9,22**
SD	**0,61**	1,03	**0,62**	1,58	2,25	**0,59**
Variance	**0,38**	1,07	**0,38**	2,50	5,08	**0,35**
CV	**0,08**	0,20	**0,08**	0,52	0,93	**0,03**

Fig. 1 Housekeeping genes 1, 3, and 6 are reliable and stably expressed in both tested groups (A and B) and can be used as reference gene (little variation in Ct/Cp values). Similar Ct/Cp values point out the similar quality of all samples

These wells, which include primers and master mix but no template, should show little or no amplification. Non-RT reactions should be performed to check for genomic DNA contamination. Amplified products should be visualized on an agarose gel under UV light.

Housekeeping Genes (HKG) should be used to control for variability in the efficiency of reverse transcription reactions [6, 7]. The housekeeping genes are genes whose expression levels do not change in the conditions being studied. One should determine cDNA quality by testing various housekeeping genes before deciding which is the most reliable as an internal control gene for the gene profiling. This additional normalization with α-tubulin, β-actin, GAPDH/G3PDH, phospholipase A2, β-2-microglobulin, or ribosomal unit 18S, 7S, etc. allows estimating the differences in quality of single samples and choosing the appropriate internal control gene. Geometric mean (GM), arithmetic mean (AM), minimal value, maximal value, standard deviation (SD), variance, and coefficient of variance (CV) should be calculated (Fig. 1).

3.5 Primer Design

Designing primers is crucial for reliable real-time PCR experiments [8]. If primer dimers or other nonspecific PCR products form, they will incorporate SYBR Green dye. This may lead to inaccurate quantification, especially when detecting sequences of low abundance. Several programs are available to help with the design including free online system: Primer3 for designing and Oligo Analyzer for controlling of primer dimers or secondary structure forming and melting temperature.

1. Designed primers should give short product (amplicon). For SYBR Green 200–300 base pair products are ideal. Template regions with secondary structures or long runs of the same nucleotide should be avoided as well as 3′ terminal Thymidine on primers.
2. Primers should have similar melting temperature.
3. Maintain a CG content of 50–60 %.
4. Primers should have low or no self-complementarity (to avoid primer dimers).
5. Primers should span exon-exon boundaries. In this way, only cDNA from mRNA gene transcripts will be amplified, not genomic copies of the gene. It is important to be aware that such primers may detect only some of the splice variants but not the other.
6. Perform in silico specificity screen; BLAST the primer sequences by running a similarity search of primers against whole organism DNA-databases to determine if primers might anneal to other targets.

3.6 PCR Efficiency

1. Make tenfold serial dilution of cDNA and run the RT-PCR reaction.
2. Plot the Ct/Cp (*y*-axis) versus log cDNA dilution (*x*-axis) and determine the slope of the line. Ideally, the dilution series will produce amplification curves that are identically spaced, doubling of product in each amplification cycle. The spacing of the curves will be determined by the equation $2^n = 10$ (dilution factor). Therefore, $n = 3.32$, and the Ct/Cp values should be separated by 3.32 cycles.
3. Amplification efficiency, *E*, is calculated from the slope of the standard curve:

$$\text{Efficiency} = 10^{-1/\text{slope}} \text{ or } \% \text{ Efficiency} = (E - 1) \quad 100\ \%$$

Efficiency close to 100 % is the best indicator of a reproducible assay. Low/too high reaction efficiencies may be caused by poor primer design or by suboptimal reaction conditions (contamination with PCR inhibitors), pipetting error in your serial dilutions or co-amplification of nonspecific products, such as primer dimers [9].

3.7 Quantifications Methods (Fig. 2)

Absolute quantification is achieved by comparing the Ct/Cp values of the test samples to a standard curve. It analyzes the quantity of nucleic acid in a given amount of sample, for example number of viral particles in certain amount of tissue or blood. Relative quantification analyzes the ratio, fold difference of a target sequence for equivalent amount of control sample (often referred as calibrator). It is important that during analyzing of samples the equivalent amounts of samples are used (previously mentioned normalization steps—weight of tissue/volume of tissue/number of cells, amount of RNA used for cDNA synthesis, and the expression level of a reference gene). For gene expression profiling relative quantification is used, because it analyzes the fold difference between control tissue, intact tissue and investigated (infected/injured) tissue. After extracting RNA from same volume of tissue and synthesizing cDNA from same amount of RNA, real-time PCR should be performed. Correctly chosen housekeeping gene (level of HKG mRNA expression is identical in control- and investigated tissue) allows determining the relative expression of target sequence by taking the ratio of the HKG-normalized target gene expression in the two investigated groups. Using only a unit mass (such as cell number, tissue volume, or μg of nucleic acid) rather than a reference gene as the normalizer can simplify the calculation. It is however strongly recommended to use all normalization steps for accurate quantification. It is possible that some differences in material quality appear

Log copy no.	1	2	3	4	5	6	7
Ct/Cp value	30,91	26,89	23,78	19,99	16,92	13,25	10,05

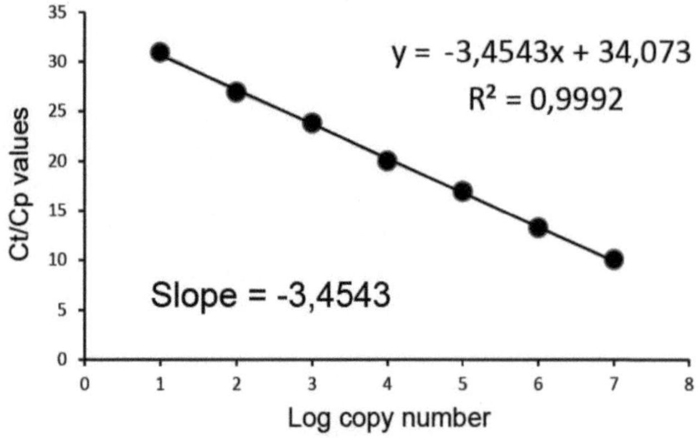

$$E = 10^{(-1/-3,4543)} = 1,948$$

$$\%E = (1,948 - 1) \times 100\% = 94,8\%$$

Fig. 2 Calculation of efficiency

during RNA extraction and cDNA synthesis. When comparing multiple samples using relative quantification, one of the samples is usually chosen as the calibrator (normal intact tissue, uninfected cells, etc.) and the expression of the target gene in all other samples is expressed as a fold change relative to the calibrator. The Ct/Cp values of target genes and reference/HKG gene (normalizer) need to be determined. Various methods can be used to determine the expression level of the target gene.

1. The Livak method, also known as the 2⁻ΔCT method assumes that both target and reference genes are amplified with efficiencies near 100 %. It is essential to verify the amplification efficiencies of the target and the reference genes and make sure that efficiencies are near 100 %.

$$\Delta CT_{(test)} = \Delta CT_{(target,test)} - \Delta CT_{(ref,test)}$$

$$\Delta CT_{(calibrator)} = \Delta CT_{(target,calibrator)} - \Delta CT_{(ref,calibrator)}$$

$$\Delta\Delta CT = \Delta CT_{(test)} - \Delta CT_{(calibrator)}$$

$$2^{-\Delta\Delta CT} = \text{Normalized expression ratio}$$

   The modified Livak method may be used if the target and the reference genes have identical amplification efficiency, but the efficiency is not equal to 2. The 2 in the equation should be replaced by the actual amplification efficiency [10].

2. The ΔCT method using a reference gene is a variation of the Livak method that gives the same results but is simpler to calculate. This method uses the difference between reference and target Ct/Cp values for each sample. Expression values must be divided by the expression value of a chosen calibrator to obtain a fold induction.

$$\text{Ratio}\,(\text{reference/target}) = 2^{CT(\text{reference})-CT(\text{target})}$$

3. The Pfaffl method must be used if the target and the reference genes do not have similar amplification efficiencies (*E*) [11].

$$\text{Ratio} = \frac{\left(E_{\text{target}}\right)^{\Delta CT,\text{target}(\text{calibrator-test})}}{\left(E_{\text{ref}}\right)^{\Delta CT,\text{ref}(\text{calibrator-test})}}$$

4 Notes

1. Upon extraction from the animal it is recommended to snap freeze the tissue in liquid nitrogen and store the samples in −80 °C until the homogenization procedure is ready to be performed. Other and often better possibility is to slice the tissue into pieces (1–3 mm wide) and keep them in RNA-later solution in a volume which is at least ten times the volume of the tissue.

In RNA-later solution RNA stay stable in room temperature up to 24 h and in 4 °C up to 7 days.

2. Formaldehyde is toxic through skin contact and inhalation of vapors; ethidium bromide is mutagenic; manipulations involving formaldehyde should be done in a chemical fume hood.
3. Extracted RNA should be devoid of contamination such as salt, protein, or genomic DNA. Poor quality RNA will affect data quality. It is crucial that the both OD A260/A280 and A260/A230 ratio are close to 2.0. The concentration of RNA should be determined by measuring the absorbance at 260 nm (A260). To ensure significance, readings should be greater than 0.15. Concentration of RNA sample = 40 × A260 × dilution factor.
4. SYBR Green is the simplest and cheapest dye used in real-time PCR technique. It binds to the double-stranded DNA and fluoresces stronger in the DNA-bound form than when unbound. Various companies provide ready to use SYBR Green master mixes which contain the dye, buffer, dNTPs, and enzyme needed for the real-time reaction.

References

1. Fasco MJ et al (1995) Quantitative RNA-polymerase chain reaction-DNA analysis by capillary electrophoresis and laser-induced fluorescence. Anal Biochem 224(1):140–147
2. Williams SJ et al (1996) Quantitative competitive polymerase chain reaction: analysis of amplified products of the HIV-1 gag gene by capillary electrophoresis with laser-induced fluorescence detection. Anal Biochem 236(1): 146–152
3. Mulder J et al (1994) Rapid and simple PCR assay for quantitation of human immunodeficiency virus type 1 RNA in plasma: application to acute retroviral infection. J Clin Microbiol 32(2):292–300
4. Heid CA et al (1996) Real time quantitative PCR. Genome Res 6(10):986–994
5. Vitzthum F et al (1999) A quantitative fluorescence-based microplate assay for the determination of double-stranded DNA using SYBR Green I and a standard ultraviolet transilluminator gel imaging system. Anal Biochem 276(1):59–64
6. Bustin SA (2000) Absolute quantification of mRNA using real-time reverse transcription polymerase chain reaction assays. J Mol Endocrinol 25(2):169–193
7. Thellin O et al (1999) Housekeeping genes as internal standards: use and limits. J Biotechnol 75(2–3):291–295
8. Li K, Brownley A (2010) Primer design for RT-PCR. Methods Mol Biol 630:271–299
9. Tichopad A et al (2003) Standardized determination of real-time PCR efficiency from a single reaction set-up. Nucleic Acids Res 31(20):e122
10. Livak KJ, Schmittgen TD (2001) Analysis of relative gene expression data using real-time quantitative PCR and the 2(-Delta Delta C(T)) Method. Methods 25(4):402–408
11. Pfaffl MW (2001) A new mathematical model for relative quantification in real-time RT-PCR. Nucleic Acids Res 29(9):e45

Chapter 14

Evaluating the Role of Nucleic Acid Antigens in Murine Models of Systemic Lupus Erythematosus

Amanda A. Watkins, Ramon G.B. Bonegio, and Ian R. Rifkin

Abstract

Impaired apoptotic cell clearance is thought to contribute to the pathogenesis of systemic autoimmune disease, in particular systemic lupus erythematosus (SLE). Endogenous RNA- and DNA-containing autoantigens released from dying cells can engage Toll-like receptors (TLR) 7/8 and TLR9, respectively in a number of immune cell types, thereby promoting innate and adaptive immune responses. Mouse models of lupus reliably phenocopy many of the characteristic features of SLE in humans and these models have proved invaluable in defining disease mechanisms. TLR7 signaling is essential for the development of autoantibodies to RNA and RNA-associated proteins like Sm and RNP, while TLR9 signaling is important for the development of antibodies to DNA and chromatin. TLR7 deficiency ameliorates end-organ disease, but, surprisingly, TLR9 deficiency exacerbates disease, possibly as a result of TLR7 overactivity in TLR9-deficient mice. Deficiency of interferon regulatory factor 5 (IRF5) inhibits autoantibody production and ameliorates disease likely due to its role in both TLR7 and TLR9 signaling. In this report we describe methods to analyze two commonly used mouse models of SLE in which TLRs and/or IRF5 have been shown to play a role in disease pathogenesis.

Key words DNA autoantigens, RNA autoantigens, Anti-nuclear autoantibody assays, Mouse models, Glomerulonephritis, Toll-like receptors, Systemic lupus erythematosus

1 Introduction

Systemic lupus erythematosus (SLE or lupus) is a chronic, relapsing and remitting autoimmune disease of unknown etiology that is characterized by the inappropriate production of inflammatory cytokines and autoantibodies of diverse specificities many of which bind to nucleic acids or their associated proteins. In patients with SLE immune-mediated injury may affect virtually any organ system including the skin, kidneys, joints, blood cells, gastrointestinal tract, heart, lungs, and brain. Multiple genetic and environmental factors interact to contribute to disease by promoting the activation of innate and adaptive immune cells. Genetic studies in humans have helped to identify genes associated with SLE susceptibility which include genes linked to TLR signaling pathways and

Hans-Joachim Anders and Adriana Migliorini (eds.), *Innate DNA and RNA Recognition*, Methods in Molecular Biology, vol. 1169, DOI 10.1007/978-1-4939-0882-0_14,

Table 1
TLR-dependent murine models of SLE discussed in this review

	MRL/*lpr*	FcγRIIB−/−	Yaa.FcγRIIB−/−
Background	MRL	C57BL/6	C57BL/6
Genetic mutations	Fas(lpr) Chr.5	FcγRIIB−/− Chr.1	FcγRIIB−/− Chr.1
			Yaa Chr. Y
50 % Mortality (month)	4(F)-6(M)	10 (F)	5 (M)
Glomerulonephritis	+++	+++	++++
Dermatitis	+++	−	−
Renal vasculitis	++	++	+++
Arthritis	+	−	−
Anti-DNA antibody	+++	+++	++
Anti-RNA antibody[a]	++	++	+++
Anti-IgG antibody	+	−	−

F female, *M* male, *Chr* chromosome, 1+ to 4+ represents titer of autoantibodies or severity of disease ranging from mild to severe
[a]Present in a subset of mice

apoptotic cell or immune complex clearance, while rodent models of SLE have proved to be invaluable tools in developing a more detailed understanding of the mechanisms of disease initiation and pathobiology of progression [1].

A number of mouse lines develop a lupus-like phenotype as a result of genetic modifications [2–4]. Several characteristics of human lupus are present in these models including production of autoantibodies specific to RNA and DNA autoantigens and the subsequent development of circulating immune complexes, splenomegaly and lymphadenopathy, and autoimmune end-organ damage including the development of dermatitis and proliferative glomerulonephritis (Table 1). These lupus-prone mice have been used to define the role of Toll-like receptors (TLRs) and their downstream signaling pathways during the initiation and effector phases of disease. We have used MRL/lpr [5, 6] and FcγRIIB knockout (FcγRIIB−/−) [4] mice in our experiments and will describe methods to analyze these two lupus-prone lines in this review.

1.1 The MRL/lpr Lupus Model

On the autoimmune-prone MRL background, mice with a mutation in the Fas gene develop lymphoproliferation (lpr) and a spontaneous lupus-like phenotype (*see* Table 1). Autoimmunity in MRL/lpr mice is due to a number of factors including an increased number of autoreactive B- and T-cells, impaired apoptosis, and an increased level of circulating, immunostimulatory RNA and DNA containing immune complexes [2, 7]. MRL/lpr mice develop autoantibodies to chromatin that are detectable with standard assays (see below) after 3-months of age and a subset of the mice also develop autoantibodies against RNA and RNA-associated proteins. Subsequently, animals develop massive adenopathy and splenomegaly, dermatitis, and glomerulonephritis. Death, which is usually due to renal disease, occurs at about 5 months in females and about 8 months in males (Table 1), although mice often have to be euthanized earlier due to severe dermatitis [5, 7, 8] (*see* **Note 1** for information on commercially available mouse strains).

1.2 Toll-Like Receptor Signaling Alters the Phenotype in MRL/lpr Mice

There is now considerable evidence to suggest that endogenous nucleic acids that are derived from injured or dying cells can be immunostimulatory via engagement of endosomal TLRs [1, 9]. TLRs that can be activated by endogenous DNA or RNA are of particular interest in lupus given that immune complexes containing these autoantigens are a hallmark of the disease. MRL/lpr mice deficient in TLR9, a receptor for DNA [10], had reduced levels of anti-DNA autoantibodies but, unexpectedly, were found to have an accelerated lupus phenotype [11, 12]. In contrast, MRL/lpr mice deficient in TLR7, a receptor for ssRNA [13, 14], developed anti-DNA autoantibodies but not anti-RNA or anti-SmRNP antibodies and exhibited reduced disease severity [11]. This highlights the important role of endogenous TLR7 signaling during disease development.

In addition to endogenous TLR ligands, pathogen-derived nucleic acids may also contribute to SLE pathogenesis. Lupus flares may occur in the setting of viral or bacterial infection. The pathogen-derived exogenous nucleic acids may engage endosomal TLRs such as TLR3 (dsRNA), TLR7 and 8 (ssRNA), and TLR9 (CpG DNA) that are expressed by many pathogenic immune cells and also by TLR-expressing parenchymal cells within the kidney, joints, and skin [15–19]. In MRL/lpr mice activation of immune cells by injection of synthetic or pathogen-derived nucleic acids that engage TLR3, TLR7, or TLR9 accelerates the development of the lupus phenotype as evidenced by earlier development of autoantibodies, proteinuria, and nephritis [15, 16, 20]. These data suggest that lupus may be accelerated by pathogen-derived nucleic acids that engage endosomal TLR3, 7, or 9.

1.3 The FcγRIIB-Deficient Lupus Model

Genetic studies in humans and mice have identified an orthologous region of chromosome 1, termed Sle1, that confers susceptibility to autoantibody production [3]. Although multiple candidate genes are located in this genetic interval, the gene encoding the inhibitory Fc gamma-receptor IIB or FcγRIIB is thought to be an important disease modifier [4, 21, 22] because FcγRIIB restrains B cell and myeloid cell activation following B cell receptor or activating Fcγ receptor engagement [23]. FcγRIIB-knockout (FcγRIIB−/−) mice that lack this inhibitory Fc-receptor develop an autoimmune disease characterized by anti-RNA and anti-DNA autoantibodies, splenomegaly and lymphadenopathy, and severe, fatal proliferative glomerulonephritis [4, 24] (Table 1). This phenotype is in part attributable to the deletion of FcγRIIB; however, it is important to recognize that more recent evidence suggests that the FcγRIIB mutation alone is not sufficient to produce full-blown disease and that additional genetic modifiers (either linked to this locus [22] or distant from this locus [21, 25]) are required for maximal disease expression (*see* **Note 2** for details).

1.4 TLR7 or IRF5 Deficiency Ameliorates Disease in FcγRIIB−/− and Yaa. FcγRIIB−/− Mice

The gene for TLR7 is located on the X-chromosome, and, due to random X inactivation, only a single copy of TLR7 is thought to be expressed in a cell. The Y-chromosome-linked autoimmune accelerator (Yaa) is a 4 Mb region of the mouse X-chromosome that spontaneously translocated to the Y chromosome and was found to potentiate autoimmunity [2]. It has recently been shown that TLR7 is the major [25, 26], but not the only [27], gene within Yaa responsible for this enhancement of autoimmunity. Yaa exacerbates autoimmunity in FcγRIIB-deficient mice [21, 22].

Interferon regulatory factor 5 (IRF5) is a transcription factor downstream of a number of TLRs including TLR7 and regulates the expression of type 1 interferons and other pro-inflammatory cytokines [28–30]. Gain-of-function IRF5 polymorphisms have been linked to lupus susceptibility in humans [31] and we have recently shown that autoimmune disease in Yaa.FcγRIIB−/− male and FcγRIIB−/− female mice is dependent on the presence of two functional copies of the IRF5 gene. IRF5-deficiency markedly decreased autoantibody production, prevented end-organ disease and prolonged survival [24]. IRF5-deficiency also led to reduced disease severity in the MRL-lpr model [32] (*see* **Note 3**).

2 Materials

2.1 Measurement of Serum Autoantibodies

2.1.1 Detection of Anti-nuclear Antibody (ANA)

1. Human Epithelial (HEp-2) cell-coated slides are used for the ANA assay. HEp-2 cells that are rapidly dividing and designed to express high levels of RNP proteins are grown on glass slides, fixed, and then permeabilized during the exponential phase of growth. The cells are in 10–12 wells per slide surrounded by a hydrophobic material that facilitates staining with as little as 20 μL of diluted serum per well.
2. Wet box for slide staining (a plastic container that can be sealed with wet paper towel inside will work fine).
3. Blocking buffer—0.2 % Bovine serum albumin (BSA) in phosphate buffered saline (PBS).

4. A fluorescently conjugated secondary antibody to mouse IgG and an epifluorescent microscope are used to detect mouse autoantibodies that have bound to DNA, RNA, or nucleic-acid associated proteins within HEp-2 cell nuclei on the slide.

2.1.2 Detection of Anti-double-stranded (ds) DNA Autoantibodies

1. Glass slides coated with *Crithidia luciliae* are used for detection of anti-dsDNA antibody titers. In contrast to the nucleus which contains a mixture of DNA, RNA, and associated proteins, the kinetoplast of *Crithidia luciliae* contains concentrated double-stranded DNA which serves as an antigen in this assay. Similar to the ANA assay, fluorescently labeled secondary antibodies and an epifluorescent microscope are required for the detection of anti-dsDNA antibodies that bind to the kinetoplast.
2. Blocking buffer—0.2 %BSA in PBS.
3. Wet box for slide staining (a plastic container that can be sealed with wet paper towel inside will work fine).
4. A fluorescently conjugated secondary antibody to mouse IgG and an epifluorescent microscope are used to detect mouse autoantibodies that have bound to dsDNA.

2.1.3 Detection of Anti-Smith/Ribonucleoprotein (Sm/RNP) Autoantibodies

1. Sm/RNP Antigen (Arotec Diagnostics Limited, Wellington, New Zealand).
2. Blocking buffer: 1 % BSA with 5 % nonfat dry milk and 0.1 % Tween 20 in PBS.
3. Wash buffer: 0.05 % Tween20 in PBS.
4. HRP-conjugated anti-mouse IgG (Sigma).
5. TMB Substrate A and B (BD Biosciences).
6. 1 M H_2SO_4 (2 N H_2SO_4).

2.2 Measurement of Kidney Disease Severity

2.2.1 Staining of Formalin-Fixed Kidney Sections

1. 10 % neutral buffered formalin for kidney fixation.
2. Clearify® Xylene alternative (American MasterTech, Lodi, CA), graded alcohols, paraffin, and cassettes for embedding of kidneys after fixation.
3. Microtome for cutting of paraffin-embedded kidney sections.
4. Hematoxylin and eosin stains.
5. Light microscope fitted with a 20 and 40× objective.

2.2.2 Staining and Cutting of Snap-Frozen Kidney Sections to Detect IgG and Complement

1. Optimal cutting temperature (OCT) embedding reagent (and cassettes for snap freezing).
2. Disposable plastic base molds.
3. 2-Methylbutane.
4. 100 % Methanol and acetone.

5. Cryostat for cutting of OCT snap-frozen kidney sections.
6. Plus glass cover slides.
7. Blocking buffer—5 % Fetal Bovine serum in PBS.
8. FITC-conjugated anti-mouse C3 (Cappel—used at a dilution of 1:150 in blocking buffer.
9. Goat anti-mouse IgG conjugated to Cy3 (Sigma)—used at a dilution of 1:100 in blocking buffer.
10. Epifluorescent microscope fitted with a 40× objective and a digital camera.

3 Methods

We use the following measures of disease severity in lupus-prone mice:

1. Lymphoid organ size and determination of the immune cell constituents.
2. Assays of serum autoantibodies and the determination of their antigen specificity.
3. Characterization of the nature and severity of kidney disease by scoring the extent of inflammation, immune complex deposition, and resulting tissue damage within the glomeruli as well as surrounding tubulo-interstitium.

3.1 Measurement of Lymphadenopathy, Splenomegaly, and Immune Cell Numbers

MRL.lpr mice and FcγRIIB−/− mice begin to show observable signs of lymphadenopathy at around 12 weeks of age. We typically measure lymph node and spleen size following euthanasia at 19 weeks of age in MRL/lpr mice, at 5 months of age in Yaa. FcγRIIB−/− male mice and at 8 months of age in FcγRIIB−/− females. We remove the spleen and cervical lymph nodes from a euthanized mouse and weigh them prior to preparation for flow cytometry if required. Analysis of the splenic or lymph node immune cell populations is then performed using flow cytometry.

3.2 Measurement of Serum Autoantibodies and the Determination of Antigen-Specificity

Carry out the following procedures at room temperature.

3.2.1 Anti-nuclear Autoantibody (ANA) Assay (See Fig. 1)

1. Serially dilute mouse serum in blocking buffer (0.2 % BSA in PBS) starting at a dilution of 1:100 and increasing in a semi-logarithmic manner (i.e., 1:100, 1:300, 1:1,000, 1:3,000, 1:10,000, 1:30,000).
2. Prepare positive and negative control sera at a 1:100 dilution. Positive control sera is from a known anti-dsDNA positive autoimmune mouse, while negative control sera is from a non-autoimmune strain such as BALB/c or C57BL/6.
3. Add 20–25 μL of diluted sera onto each well of a HEp2 slide making sure to cover the entire well area. The hydrophobic

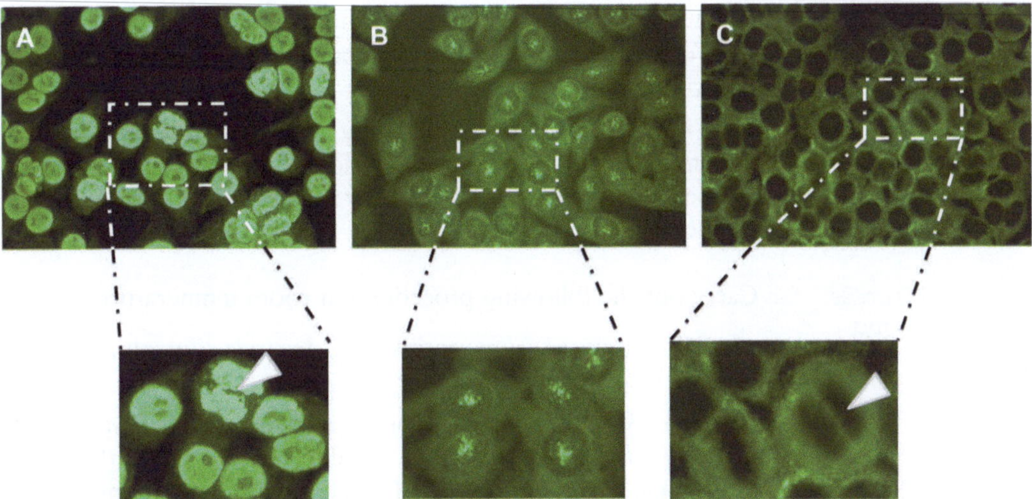

Fig. 1 *Anti-nuclear autoantibody (ANA) assay.* Serum from (**a**) an MRL.lpr mouse, (**b**) a male Yaa.FcγRIIB–/– mouse, and (**c**) a female FcγRIIB–/– mouse deficient in IRF5 was diluted 1:100 and incubated on HEp2 slides. Subsequently, slides were incubated with a goat anti-mouse IgG antibody conjugated to FITC and imaged at 200× magnification using a fluorescent microscope. A homogeneous nuclear staining pattern (**a**) characteristic of MRL.lpr mice in which the nucleus is diffusely stained. The magnified insert shows the positive staining of chromatin in a dividing cell (*arrowhead*) indicative of anti-chromatin autoantibodies. Discrete speckled staining (**b**) is more suggestive of anti-RNA autoantibodies in the serum of Yaa.FcγRIIB–/– mice. ANA staining is negative in serum from the FcγRIIB mouse deficient in IRF5 (**c**) thus anaphase chromosomes are also negative (*arrowhead* in **c**). The positive staining in (**c**) represents cytoplasmic staining (*see* **Notes 4** and **5**)

area around the well should remain dry while the diluted sera should form a drop and remain in the well.

4. Incubate the slides at room temperature for 1 h in a moist chamber.
5. Wash the slides once by pipetting 25 mL of PBS directly onto the HEp2 slide. Then wash the slides 3×5 min in fresh PBS within a Coplin jar. Agitating the jar a bit may reduce back ground staining.
6. Blot the slides on paper towel and carefully dry the hydrophobic regions around the wells with Kimwipe. Do not let the wells become dry.
7. Dilute 2.5 μL of the fluorophore-conjugated goat anti-mouse antibody in 247.5 μL of blocking buffer (i.e., 1:100) for each HEp-2 slide—make a mastermix sufficient to cover all wells on all HEp2 slides being used that day for the assay.
8. Add 25 μL to each well of the HEp2 slide.
9. Incubate the slides for 1 h in a moist chamber at room temperature in the dark.
10. Repeat **steps 4** and **5**.

11. Add a few drops of aqueous mounting media (VECTASHIELD, Vector Laboratories, Burlingame, CA) to the slides and cover with glass coverslips.
12. View and/or photograph using an epifluorescent microscope (*see* **Notes 4** and **5** for information on analysis). The slides may be stored at 4 °C in the dark for several days prior to viewing if necessary.

3.2.2 *Crithidia luciliae* Assay for Anti-dsDNA Autoantiboies (See Fig. 2)

Carry out the following procedures at room temperature.

1. Dilute mouse serum, positive and negative controls at 1:100 in blocking buffer.
2. Add 25 μL of diluted sera onto each well of a *Crithidia luciliae* slide making sure to cover the entire well area. If the hydrophobic area around the well is dry, the diluted sera should form a drop and remain in the well.
3. Incubate the slides for 1 h in a moist chamber
4. Initially wash the slide by pipetting 25 mL of PBS directly onto the slide. Then wash the slides 3 × 5 min in fresh PBS within a Coplin jar.
5. Blot the slides on paper towel and dry the hydrophobic regions around the wells with a Kimwipe. Do not let the wells become completely dry.
6. Dilute anti-mouse IgG antibody conjugated to a fluorophore 1:100 in blocking buffer making a mastermix sufficient to cover all wells on all the slides being used that day for the assay (250 μL/slide or 25 μL/well).
7. Add 25 μL to each well of the Crithidia slide and incubate the slides for 1 h in a dark, moist chamber.

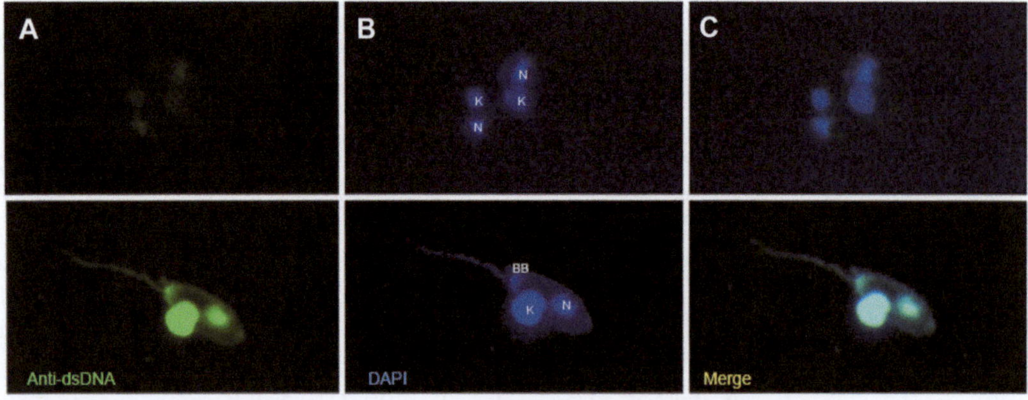

Fig. 2 *Crithidia luciliae assay.* Crithidia slides were incubated with serum from a C57BL/6 control mouse (**a**) and from an anti-dsDNA positive Yaa.FcγRIIB−/− lupus-prone mouse **(b)**. Anti-dsDNA antibodies were detected using an anti-mouse antibody conjugated to FITC (*left hand panels,* a), DAPI was used to counterstain the nucleus (N), kinetoplast (K), and basal body (BB) (*center panels,* b). The merged images (*right hand panels,* **c**) show strong colocalization of anti-dsDNA antibodies in the kineto plast

8. Repeat washes as described in **steps 4** and **5**.
9. Add a few drops of VECTASHIELD mounting media (containing DAPI to identify nucleus and kinetoplast) to the slides and cover with glass coverslips. Store the slides at 4 °C in the dark until ready to view (*see* **Note 6** for information on analysis).

3.2.3 Enzyme-Linked Immunosorbent Assay (ELISA) for Anti-Sm/RNP Antibodies

Carry out the following procedures at room temperature unless otherwise noted.

1. Coat half of a 96-well ELISA plate (i.e., 48 wells) with Sm/RNP antigen at 2 μg/mL in a volume of 50 μL/well. As a negative control coat the remaining wells with BSA at a concentration of 2 μg/mL in a volume of 50 μL/well (*see* **Note 7**).
2. Cover the plate with Parafilm and incubate at 4 °C overnight.
3. Wash three times with wash buffer.
4. Block for 2 h at room temperature using 200 μL/well of blocking buffer.
5. Wash three times with wash buffer (0.05 % Tween20 in PBS).
6. Prepare a standard curve with an anti-Sm/RNP antibody by performing half-log serial dilutions (1:3.33) of the antibody in blocking buffer (we start at 1 μg/mL of an anti-Sm/RNP monoclonal and dilute to 2 ng/ml).
7. Dilute experimental serum samples 1:100 in blocking buffer (make at least 250 μL/sample).
8. Add each sample to two Sm/RNP-coated wells (50 μL/well) and two BSA-coated wells (50 μL/well).
9. Incubate for 1 h at room temperature.
10. Wash 5 times in wash buffer.
11. Detect bound IgG by incubating for 1 h using HRP-conjugated anti-mouse IgG 1:3,000 in blocking buffer.
12. Wash 3 times in wash buffer.
13. Develop using TMB substrate (BD Biosciences).

3.3 Measurement of Kidney Disease Severity

3.3.1 Scoring Kidney Disease Severity by Light Microscopy (See Fig. 3)

1. Euthanize the mouse.
2. Prior to harvesting the kidneys, perfuse the mouse by injecting 20 mL of phosphate buffered saline via the left ventricle to remove red blood cells from kidney.
3. Remove the kidney capsule from both kidneys using a scalpel and forceps.
4. Slice each kidney into three sections; an upper and lower pole section and a mid-pole section.
5. The upper and lower poles of each kidney can be used for analysis of RNA and protein or for flow cytometry.

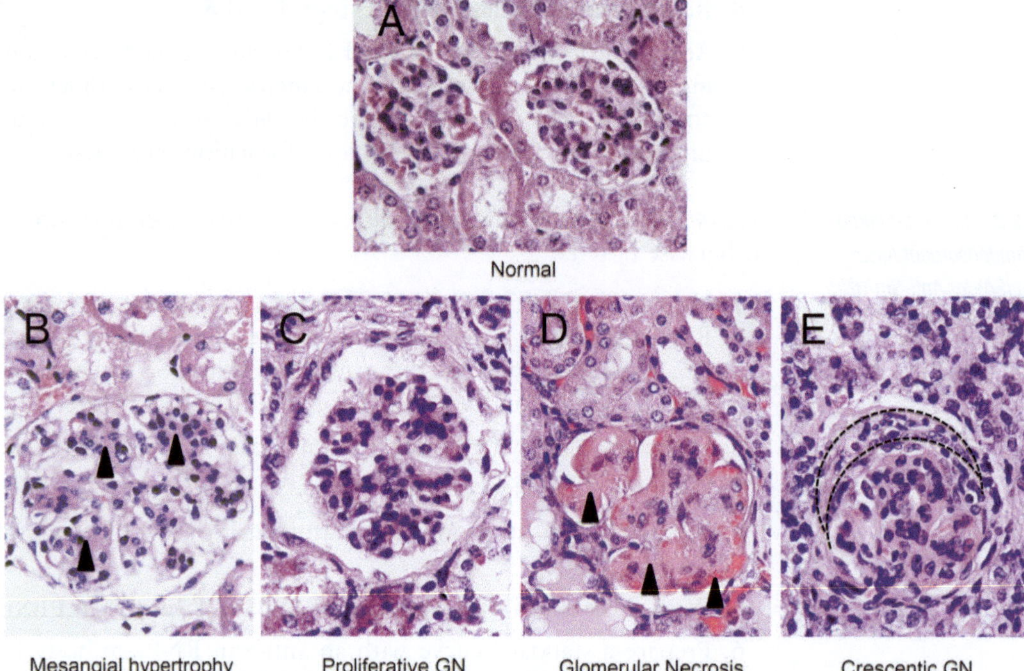

Fig. 3 *Scoring glomerular disease severity in lupus-prone mice.* Formalin-fixed kidneys are scored based on the analysis of 50 glomeruli from each mouse. Each glomerulus is examined and given a score from 0 (**a**) through 4 (**e**). Normal glomeruli (**a**) have a fine lace appearance to the mesangium and capillary loops and usually have about 30–40 nuclei depending on the size of the glomerulus and the thickness of the section. Mild glomerular injury (**b**) results in mesangial hypertrophy (*arrowheads*). Infiltration of the glomerulus by inflammatory cells or proliferation of resident glomerular cells results in a "proliferative lesion" (**c**) that can be recognized by an increase in the number of nuclei present in the glomerular tuft. Severe injury (**d**, **e**) causes tuft necrosis (*arrowheads* in **d**) or extracapillary crescent formation (*outlined* in **e**)

6. Place mid-pole (2–3 mm thick) into a tissue cassette.
7. Immediately place cassette in 10 % neutral buffered formalin and fix tissue overnight at room temperature.
8. Process and paraffin-embed the tissue using standard techniques.
9. Use a microtome to cut 8 μm thick paraffin sections and then stain using hematoxylin and eosin (H and E) or Periodic Acid Shiff (PAS).
10. An investigator blinded to sample identity then grades the extent of glomerular and interstitial disease as described in steps 11–12 below.
11. Twenty-five to fifty glomeruli in one or two different kidney sections from each mouse are examined and the severity of injury in each glomerulus is measured using a scale of 0-4 (Fig. 3) where 0 = Normal, 1 = Mesangial Hypertrophy, 2 = Proliferative Glomerulonephritis and/or immune cell infiltration, 3 = Glomerular Necrosis and 4 = Crescentic Glomerulonephritis;

mean glomerular score is the sum of all the individual glomerular scores from a particular mouse divided by the total number of glomeruli examined (*see* **Notes 8** and **9**).

12. Interstitial disease (*see* **Note 10**) is evaluated by examining 10–20 randomly selected, low-powered fields of kidney cortex and scoring the amount of immune cell infiltrate and fibrosis on a scale from 0 = normal to 3 = >75 % of the interstitium is inflamed or fibrosed: mean interstitial score in an individual mouse is the sum of scores of all the selected fields of kidney cortex in that particular mouse divided by the total number of fields examined (*see* **Note 11**).

3.3.2 Quantification of Glomerular IgG and Complement Deposition by Immunofluorescence (See Fig. 4)

1. Place a beaker of 2-methylbutane in a container of dry ice and place TissueTek OCT into disposable plastic molds.
2. Harvest the kidney as previously described in Subheading 3.3.1.
3. Immediately place a 2–3 mm section of the mid-pole of the kidney into the mold containing OCT and cover tissue completely with additional OCT (avoid getting bubbles into the OCT).
4. Snap-freeze the samples by gently lowering them into the cold methylbutane using long forceps, taking care not to let the molds fall over during freezing.
5. After the kidneys are frozen in OCT, wrap them in aluminum foil and store them at −80 °C until they are sectioned.
6. Precool the cryostat to −22 °C and place the blocks (kidneys frozen in OCT) to be cut into the chamber to equilibrate with the cryostat temperature.
7. Cut 5 μm thick sections and place 2–3 sections onto each glass slide (*see* **Note 12**).
8. Place cut slides into a slide box inside the cryostat (i.e., at −22 °C), seal the slide box with Parafilm and transfer the slides to −80 °C until they are stained.

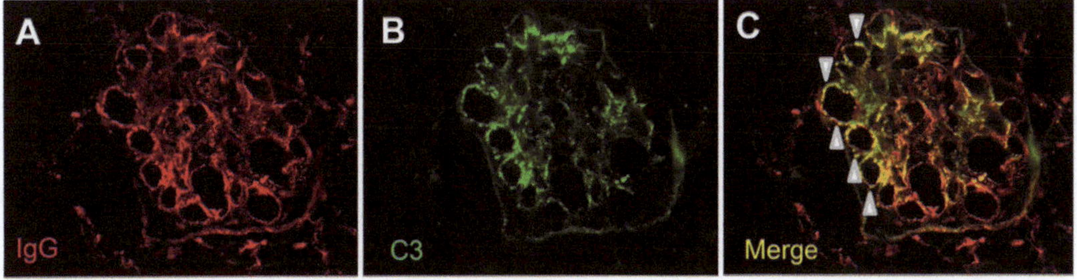

Fig. 4 IgG and complement deposition in kidney glomeruli. A glomerulus from a FcγRIIB−/− female mouse was stained using an anti-mouse IgG antibody (**a**) and an anti-mouse C3 antibody (**b**). Immune complexes containing IgG associated with C3 can be observed in the merged image (**c**). Granular appearing immune-complexes are present in the glomerular capillary loops (arrowheads)

9. Prior to fixation and staining, pre-chill a 1:1 v/v methanol–acetone mixture to −20 °C.
10. Wash the slides briefly in PBS to remove excess OCT and then place in methanol–acetone at −20 °C for 10 min to fix the tissue.
11. Wash the slides 3×5 min with 1× PBS at room temperature and then tap the slides on paper towel to remove excess PBS.
12. Circle the tissue section using a hydrophobic PAP pen.
13. To block, put 100 μL of blocking buffer (1 % BSA in PBS) on each tissue section and incubate in a moist chamber for 30 min at room temperature.
14. Prepare the appropriate dilution of fluorescent-labeled anti-IgG and/or anti-C3 antibody in 1 % BSA in PBS, keeping the antibodies on ice and in the dark.
15. Carefully suction off the blocking buffer and quickly replace with 100 μL of diluted antibody.
16. Incubate in a dark, moist chamber for 90 min at 25 °C at room temperature.
17. Wash the slides 3×5 min with 1× PBS keeping the slides in the dark as much as possible.
18. Tap the slides dry on paper towel and suction away any excess PBS.
19. Place VECTASHIELD on the tissue sections, cover with a coverslip, seal with nail polish, and allow the slides to dry in the dark before microscopy (the slides can be kept in the dark at 4 °C for up to 1 week before imaging, but do not freeze).
20. Image 20 glomeruli from each mouse using an epifluorescent microscope fitted with a digital camera—this should be done by an investigator blinded to sample identity.
21. Quantify the fluorescence intensity of each glomerulus stained with either anti-IgG or anti-C3 using Image J or Adobe Photoshop software.
22. Calculate the median fluorescence intensity of the 20 imaged glomeruli for both anti-IgG and anti-C3 stains (*see* **Note 13**).

4 Notes

1. Two MRL/lpr mouse strains are commercially available from Jackson Laboratories. As a result of genetic drift the colony of MRL/lpr mice at Jackson (Stock #006825) was noted in 2006 to have a progressive loss of the lymphoproliferative and lupus-like phenotype, and this was associated with an increased life span. In 2007 the MRL/lpr line, which is now designated MRL/MpJ-Faslpr/J (Stock #000485) was recovered from cryopreserved embryos. The recovered line has a phenotype

similar to that previously described by Kelley and Roths in 1985 [7] and develops severe T-cell proliferation, arthritis, immune complex glomerulonephritis and dies between 17 and 22 weeks. We have preferred to use this strain in our recent experiments.

2. The gene for FcγRIIB lies within the Sle1 susceptibility locus and deficiency of this inhibitory Fc-receptor promotes the development of immune complex disease [4]. However, it has recently been recognized that other genes in this region also influence the autoimmune phenotype [22, 33]. This should be considered when evaluating the lupus phenotype of knockout strains that are bred to FcγRIIB-deficient mice.

3. It has been recently recognized that the IRF5−/− line (on the C57BL/6 background) had an additional mutation in the DOCK2 gene [34, 35]. This should be considered in the interpretation of data generated using IRF5−/− mice.

4. Anti-nuclear antibody staining (*see* Fig. 1): When performing an ANA assay using serum from either the Yaa.FcγRIIB−/− or MRL.lpr models of lupus, the most common staining patterns observed are homogenous with the nucleoli remaining unstained (Fig. 1a). A homogenous staining pattern typically signifies the presence of anti-ssDNA, anti-dsDNA, anti-chromatin, and/or anti-histone autoantibodies. Anti-chromatin autoantibodies can also be detected if a HEp2 cell happened to be undergoing mitosis at the time of fixation (Fig. 1a). The chromatin will be easily observed in anaphase at both poles of the dividing cell if anti-chromatin antibodies are present. A nuclear speckled (fine or course) pattern (Fig. 1b) can also be observed if antibodies to RNA or RNA-associated antigens are present: (e.g., Smith antigen (Sm), ribonucleoproteins (RNP), SSA/Ro, SSB/La, Scl-70, Jo-1, or PM1). In mice that have both anti-DNA and anti-RNA antibodies, the nuclear speckled staining may at first be masked by the homogeneous pattern; however, it may become more apparent as the serum is diluted. Cytoplasmic staining on HEp2 immunofluorescence (Fig. 1c) is incompletely understood but may be associated with anti-RNA or myositis-specific antibodies [36].

5. Recommended titrations for performing ANA using Yaa.FcγRIIB−/− or MRL.lpr serum samples are as follows: 1:100, 1:300, 1:1,000, 1:3,000, 1:10,000, 1:30,000. On each slide, a positive and negative control are needed (*see* Subheading 3.2.1). The data is scored in a blinded manner as follows: (Negative) = Little to no staining compared to negative control; (+/−) slightly more staining compared to negative control but pattern cannot be discerned; (+) clearly positive compared to negative control but not nearly as bright as the positive control diluted 1:100; (++) clearly positive and twice as bright as those

scored +; (+++) clearly positive and three times as bright as those scored +; and (++++) clearly positive and fluorescent intensity is equal if not brighter than positive control diluted 1:100. The result for each mouse is the last positive + dilution (e.g., a sample scored +++ at a dilution of 1:1,000, + at 1:3,000, and +/− at 1:10,000 should be scored as 1:3,000). If scoring the ANA slides from photographed images it is important that all images are taken using the same camera settings, i.e., constant exposure time, gain, binning, etc. The camera settings used should reflect as closely as possible what is seen by eye under the microscope. The negative control should appear very dark in images while the images of the positive control should be bright but not oversaturated.

6. *Crithidia luciliae* assay for anti-dsDNA (*see* Fig. 2). The Crithidia assay allows for the detection of autoantibodies with specificity for dsDNA which is also called native DNA (nDNA). The kinetoplast of *Crithidia luciliae* (Fig. 2) consists of dsDNA with trivial amounts of ssDNA, nucleoprotein, or RNA antigens [37]. It is recommended to counterstain with a dye (e.g., DAPI) which will stain both the nucleus as well as the kinetoplast (Fig. 2). The kinetoplast is located closer to the basal body and flagellum than the nucleus. The flagellum is more easily visible under bright field than the basal body, kinetoplast, or nucleus. Therefore, light microscopy may be helpful to visualize the slide if the kinetoplast is unstained and therefore cannot be seen using fluorescent microscope settings.
7. Sm/RNP ELISA. In order to ensure that the anti-Sm/RNP antibody is not nonspecifically binding to the Sm/RNP-coated wells, it is important to subtract the signal of the BSA-coated wells (should be close to zero) from that of the Sm/RNP-coated wells prior to calculating the concentration from the standard curve. If the optical density of a sample lies above the highest point on the standard curve, additional dilutions are required for quantification.
8. It is crucial for analysis of kidney histology for the pathologist/investigator who is scoring or photographing the sections to be blinded to the study groups and for the analysis to be performed on a large number of glomeruli for each mouse. Lupus is characteristically a patchy disease and so conscious or subconscious selection of more or less damaged area may invalidate the analysis.
9. Use of the glomerular score is categorical data, so groups should be compared using a Wilcoxon signed-rank test or an Analysis of Variance (ANOVA) on ranks.
10. Interstitial inflammation refers to infiltration of inflammatory cells and the subsequent development of fibrosis in the regions of the kidney between tubules and glomeruli.

11. Although interstitial disease appears to develop later in the course of murine lupus nephritis than glomerular changes, there is usually a good correlation between glomerular scores and interstitial scores.
12. Make sure that there are sufficient sections on each slide so that isotype control staining can be done on the same slide.
13. It is important to include kidney sections from healthy non-autoimmune mice in the analysis to demonstrate the extent to which the kidney disease seen in the experimental mice differs from normal. We include as controls healthy age-matched C57BL6 mice for studies of Yaa.FcγRIIB−/− and FcγRIIB−/− or young (4–6 week old) MRL mice for studies of MRL.*lpr* . These should be stained with the experimental samples and included in the coded slides sent for scoring.

Acknowledgements

This work was supported by a grant from the Alliance for Lupus Research (I.R.R) and the following grants from the National Institutes of Health: P01 AR050256 (I.R.R), RO1 DK090558 (R.G.B.B), and a Research Training in Immunology T32 Grant AI007309-23 (A.A.W).

References

1. Liu Z, Davidson A (2012) Taming lupus-a new understanding of pathogenesis is leading to clinical advances. Nat Med 18:871–882
2. Perry D, Sang A, Yin Y, Zheng YY, Morel L (2011) Murine models of systemic lupus erythematosus. J Biomed Biotechnol 2011: 271694
3. Cheung YH, Loh C, Pau E, Kim J, Wither J (2009) Insights into the genetic basis and immunopathogenesis of systemic lupus erythematosus from the study of mouse models. Semin Immunol 21:372–382
4. Bolland S, Ravetch JV (2000) Spontaneous autoimmune disease in Fc(gamma)RIIB-deficient mice results from strain-specific epistasis. Immunity 13:277–285
5. Andrews BS, Eisenberg RA, Theofilopoulos AN et al (1978) Spontaneous murine lupus-like syndromes. Clinical and immunopathological manifestations in several strains. J Exp Med 148:1198–1215
6. Dixon FJ, Andrews BS, Eisenberg RA, McConahey PJ, Theofilopoulos AN, Wilson CB (1978) Etiology and pathogenesis of a spontaneous lupus-like syndrome in mice. Arthritis Rheum 21:S64–S67
7. Kelley VE, Roths JB (1985) Interaction of mutant lpr gene with background strain influences renal disease. Clin Immunol Immunopathol 37:220–229
8. Ghoreishi M, Dutz JP (2009) Murine models of cutaneous involvement in lupus erythematosus. Autoimmun Rev 8:484–487
9. Marshak-Rothstein A, Rifkin IR (2007) Immunologically active autoantigens: the role of toll-like receptors in the development of chronic inflammatory disease. Annu Rev Immunol 25:419–441
10. Hemmi H, Takeuchi O, Kawai T et al (2000) A Toll-like receptor recognizes bacterial DNA. Nature 408:740–745
11. Christensen SR, Shupe J, Nickerson K, Kashgarian M, Flavell RA, Shlomchik MJ (2006) Toll-like receptor 7 and TLR9 dictate autoantibody specificity and have opposing inflammatory and regulatory roles in a murine model of lupus. Immunity 25:417–428
12. Christensen SR, Kashgarian M, Alexopoulou L, Flavell RA, Akira S, Shlomchik MJ (2005) Toll-like receptor 9 controls anti-DNA autoantibody production in murine lupus. J Exp Med 202:321–331

13. Diebold SS, Kaisho T, Hemmi H, Akira S, Reis e Sousa C (2004) Innate antiviral responses by means of TLR7-mediated recognition of single-stranded RNA. Science 303:1529–1531
14. Heil F, Hemmi H, Hochrein H et al (2004) Species-specific recognition of single-stranded RNA via toll-like receptor 7 and 8. Science 303:1526–1529
15. Patole PS, Grone HJ, Segerer S et al (2005) Viral double-stranded RNA aggravates lupus nephritis through Toll-like receptor 3 on glomerular mesangial cells and antigen-presenting cells. J Am Soc Nephrol 16:1326–1338
16. Pawar RD, Patole PS, Zecher D et al (2006) Toll-like receptor-7 modulates immune complex glomerulonephritis. J Am Soc Nephrol 17: 141–149
17. Drexler SK, Sacre SM, Foxwell BM (2006) Toll-like receptors: a new target in rheumatoid arthritis? Expert Rev Clin Immunol 2: 585–599
18. Roelofs MF, Joosten LA, Abdollahi-Roodsaz S et al (2005) The expression of toll-like receptors 3 and 7 in rheumatoid arthritis synovium is increased and costimulation of toll-like receptors 3, 4, and 7/8 results in synergistic cytokine production by dendritic cells. Arthritis Rheum 52:2313–2322
19. Miller LS, Modlin RL (2007) Toll-like receptors in the skin. Semin Immunopathol 29: 15–26
20. Anders HJ, Vielhauer V, Eis V et al (2004) Activation of toll-like receptor-9 induces progression of renal disease in MRL-Fas(lpr) mice. FASEB J 18:534–536
21. Bolland S, Yim YS, Tus K, Wakeland EK, Ravetch JV (2002) Genetic modifiers of systemic lupus erythematosus in FcgammaRIIB(−/−) mice. J Exp Med 195:1167–1174
22. Boross P, Arandhara VL, Martin-Ramirez J et al (2011) The inhibiting Fc receptor for IgG, FcgammaRIIB, is a modifier of autoimmune susceptibility. J Immunol 187:1304–1313
23. Bolland S, Ravetch JV (1999) Inhibitory pathways triggered by ITIM-containing receptors. Adv Immunol 72:149–177
24. Richez C, Yasuda K, Bonegio RG et al (2010) IFN regulatory factor 5 is required for disease development in the FcgammaRIIB−/−Yaa and FcgammaRIIB−/− mouse models of systemic lupus erythematosus. J Immunol 184:796–806
25. Subramanian S, Tus K, Li QZ et al (2006) A Tlr7 translocation accelerates systemic autoimmunity in murine lupus. Proc Natl Acad Sci U S A 103:9970–9975
26. Pisitkun P, Deane JA, Difilippantonio MJ, Tarasenko T, Satterthwaite AB, Bolland S (2006) Autoreactive B cell responses to RNA-related antigens due to TLR7 gene duplication. Science 312:1669–1672
27. Santiago-Raber ML, Kikuchi S, Borel P et al (2008) Evidence for genes in addition to Tlr7 in the Yaa translocation linked with acceleration of systemic lupus erythematosus. J Immunol 181:1556–1562
28. Savitsky D, Tamura T, Yanai H, Taniguchi T (2010) Regulation of immunity and oncogenesis by the IRF transcription factor family. Cancer Immunol Immunother 59:489–510
29. Ronnblom L (2011) The type I interferon system in the etiopathogenesis of autoimmune diseases. Ups J Med Sci 116:227–237
30. Paun A, Pitha PM (2007) The IRF family, revisited. Biochimie 89:744–753
31. Deng Y, Tsao BP (2010) Genetic susceptibility to systemic lupus erythematosus in the genomic era. Nat Rev Rheumatol 6:683–692
32. Tada Y, Kondo S, Aoki S et al (2011) Interferon regulatory factor 5 is critical for the development of lupus in MRL/lpr mice. Arthritis Rheum 63:738–748
33. Sato-Hayashizaki A, Ohtsuji M, Lin Q et al (2011) Presumptive role of 129 strain-derived Sle16 locus in rheumatoid arthritis in a new mouse model with Fcgamma receptor type IIb-deficient C57BL/6 genetic background. Arthritis Rheum 63:2930–2938
34. Purtha WE, Swiecki M, Colonna M, Diamond MS, Bhattacharya D (2012) Spontaneous mutation of the Dock2 gene in Irf5−/− mice complicates interpretation of type I interferon production and antibody responses. Proc Natl Acad Sci U S A 109:E898–E904
35. Yasuda K, Nündel K, Watkins AA et al (2013) Phenotype and function of B cells and dendritic cells from interferon regulatory factor 5-deficient mice with and without a mutation in DOCK2. Int Immunol 25:295–306
36. Wiik AS, Hoier-Madsen M, Forslid J, Charles P, Meyrowitsch J (2010) Antinuclear antibodies: a contemporary nomenclature using HEp-2 cells. J Autoimmun 35:276–290
37. Slater NG, Cameron JS, Lessof MH (1976) The Crithidia luciliae kinetoplast immunofluorescence test in systemic lupus erythematosus. Clin Exp Immunol 25:480–486

Chapter 15

Induction and Analysis of Nephrotoxic Serum Nephritis in Mice

John M. Hoppe and Volker Vielhauer

Abstract

The nephrotoxic serum nephritis (NTN) model is an integral part of experimental glomerulonephritis (GN) research. Here, we discuss how the murine NTN model can be induced and effectively used to mimic an immune complex-mediated GN. Further, we differentiate between heterologous and autologous models by comparing pathophysiology and phenotypic manifestations.

Key words NTN, Glomerulonephritis, Albuminuria, IgG, Complement, Mouse

1 Introduction

Glomerulonephritis (GN) is a renal disease with a wide spectrum of clinical manifestations and an even larger range of pathogenic mechanisms. Most forms, however, are triggered by innate and adaptive immune responses directed against glomerular structures, commonly immune complexes formed in situ or deposited in the mesangium and glomerular capillary walls. A well-established rodent model for studying the pathophysiology of immune complex-induced GN in mice and rats is nephrotoxic serum nephritis (NTN). This model does not depend on pathogenic events that are otherwise required for the formation of pathogenic autoantibodies, facilitating direct assessment of renal injury in different genotypes or treatment regimens induced by the same immunological insult (*see* **Note 1**). As early as 1900, the Russian scientist Lindeman theorized on the possibility of using anti-kidney antibodies to induce a GN. Based on this concept, in 1934 the Japanese pathologist M. Masugi was first to establish a working model of NTN, after whom the synonym Masugi nephritis was named after [1, 2]. Then in 1936 Smadel and collaborators described in detail the pathophysiology in rats of both the acute and chronic stages of the NTN [3]. Moreover, they examined strain and sex differences in rats and

Hans-Joachim Anders and Adriana Migliorini (eds.), *Innate DNA and RNA Recognition*, Methods in Molecular Biology, vol. 1169, DOI 10.1007/978-1-4939-0882-0_15, © Springer Science+Business Media New York 2014

investigated on anaphylactic effects of the nephrotoxic serum [4]. During the early 1940s, Kay was able to correlate symptoms of renal injury with the host's antibody response to the nephrotoxic serum [5]. Over the next decades it became clear that the pathogenic mechanisms in this model are relatively complex and variable. This variability is exacerbated by the abundance of NTN models applied. The focus of this article is on NTN models induced in mice with rabbit nephrotoxic serum (*see* **Notes 2** and **3**). References to other models will be specified.

Importantly, universal entities exist that all models have in common. First, their general setup is alike. An experimental animal (rat or mouse) is injected with polyclonal nephrotoxic serum containing antibodies against glomerular structures, including the glomerular basement membrane (GBM). Hence, the NTN model is sometimes also referred to as anti-GBM antibody-induced GN. The nephrotoxic serum is raised in a second animal species (commonly rabbit or sheep) after immunization with glomerular preparations of the first species. Secondly, most nephrotoxic serum preparations are crude. Besides anti-GBM antibodies, the serum contains immunoglobulins against other structural elements of the kidney. Thirdly, even though the host's inflammatory response in this model varies vastly dependent on host species, donor serum species, and serum toxicity, it can generally be divided into two separate phases:

The initial acute (heterologous) phase of inflammation (Fig. 1a, b) is triggered by glomerular deposition of heterologous antibodies contained in the nephrotoxic serum within minutes after injection. The polyclonal antibodies bind to antigens of the glomerular endothelial wall and GBM, locally triggering innate immune activation, including secretion of proinflammatory chemokines and cytokines. Antibody deposition may also directly interfere with interconnecting integrins and induce podocytes injury with foot process effacement [6]. In the heterologous phase, glomerular injury depends largely on complement activation and neutrophil-dominated inflammation [7–10]. Acute phase complement proteins bind to the conserved Fc region of nephrotoxic IgG antibodies deposited in glomeruli. Subsequently, sublytic C5a plays a major role in triggering neutrophil and monocyte chemotaxis and activation [11]. Lytic complement components appear to be less relevant for NTN pathogenesis, as self-cells are protected from membrane attack complexes [12]. Yet the extent of complement involvement in NTN is still in debate, as some studies have shown that heterologous phase damage can be complement- and neutrophil-independent [10, 13]. Nonetheless, evidence is mounting that, except for high doses of nephrotoxic serum, in the heterologous phase renal injury is complement-dependent [8, 14]. Furthermore, stimulation of Toll-like receptors (TLRs) on both bone marrow-

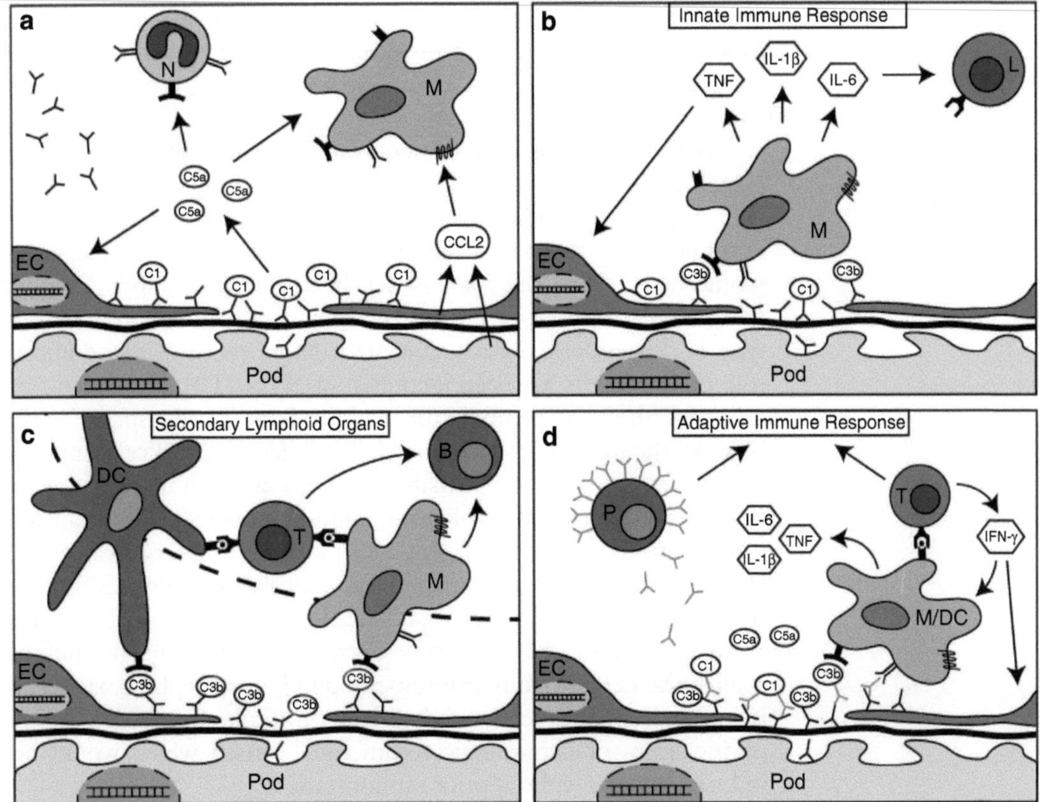

Fig. 1 Pathogenic mechanisms of nephrotoxic nephritis in the heterologous (**a**, **b**) and autologous phase (**c**, **d**). (**a**) Initial adherence of heterologous nephrotoxic antibody (*black*) to structures of the glomerular basement membrane (GBM), endothelial cells (EC), and podocytes (Pod). Formation of immune complexes activates complement components (C1, C3b, C5a), which in turn induce neutrophil (N) and monocyte (M) recruitment and activation. Sublytic complement products may also activate intrinsic glomerular cells to secrete proinflammatory chemokines like CCL2/MCP-1 and cytokines, which augment glomerular leukocyte infiltration. (**b**) Complement-opsonized heterologous antibodies are endocytosed by activated mononuclear phagocytes (M), which secrete further cytokines like IL-1β, IL-6, and TNF. These stimulate intrinsic glomerular cells, increase endothelial permeability, activate lymphocytes (L), and together with ongoing leukocyte infiltration lead to exudative and proliferative glomerular injury. (**c**) Locally and in secondary lymphatic tissue dendritic cells (DC) and macrophages (M) present heterologous nephrotoxic antibody epitopes to T lymphocytes (T), inducing the activation of nephritogenic CD4+ and CD8+ cells. Activated CD4+ T cells induce B-cell (B) maturation in lymphoid tissues. (**d**) Kidney-infiltrating nephritogenic CD4+ T cells secrete Th1 effector cytokines such as IFN-γ and further stimulate glomerular macrophages. Matured plasma cells (P) produce autologous antibodies (*light grey*) directed against the heterologous nephrotoxic antibodies planted in glomeruli. Autologous antibodies are additionally deposited in the glomerulus to form large immune complexes that further augment local inflammatory responses

derived and resident renal cells (including TLR2, TLR3, TLR4, and TLR5), as well as activation of cytosolic RNA recognition receptors like melanoma differentiation-associated gene (MDA)-5 synergistically activates renal production of chemokines and cytokines and exacerbates NTN [15–17]. Dependent on the dose and toxicity of the nephrotoxic serum applied, glomerular injury in the heterologous phase can be self-limiting or progress to capillary distortion and thrombosis, crescent formation, and glomerulosclerosis.

With a sufficient dose of nephrotoxic serum given, the chronic (autologous) phase of NTN develops within approximately 1 week (Fig. 1c, d). It is characterized by the emergence of systemic adaptive immune responses against the heterologous antibody planted in the glomerulus. A Th1 cell response has been particularly well described that stimulates macrophages to mediate renal damage in a classical type IV hypersensitivity reaction [18]. More recently, an additional role of Th17 cells has been identified in NTN [19]. Renal dendritic cells were shown to promote the nephritogenic T cell response in this phase of the model [20]. In parallel, T-cell dependent B cell activation in the host promotes the formation of autologous antibodies directed against the injected heterologous immunoglobulins. These autologous antibodies are additionally deposited in the glomerulus and exacerbate disease by forming large glomerular immune complexes containing both heterologous and autologous immunoglobulin. The autologous phase of NTN can be boostered by immunizing the host against the heterologous immunoglobulin before nephrotoxic serum is injected. In this accelerated variant of autologous NTN, subnephritic doses of nephrotoxic serum can be used which would not induce nephritis without prior immunization.

Functionally in both the heterologous and autologous model, yet to a varying extent and time course, animals show signs of renal dysfunction, including proteinuria (Fig. 2a), a nephrotic syndrome like hypoproteinemia and hyperlipoproteinemia, and renal insufficiency, as revealed by increases in serum creatinine and urea (Fig. 2b). By choosing the appropriate time frame, the heterologous phase of NTN can be studied independently of the autologous phase. We suggest that heterologous disease models should be 7 days or shorter (injection to endpoint), since autologous immunoglobulin production reaches significant levels 7–10 days after nephrotoxic serum administration.

Renal leukocyte infiltration is a hallmark of the acute and the chronic phase of NTN. Histologic and flow-cytometric analysis show that short heterologous models induce glomerular leucocyte infiltration with a characteristically large neutrophil population [7, 10]. Glomerular injury in chronic autologous models is associated with significant glomerular infiltrates of mononuclear phagocytes (macrophages, dendritic cells) and T-cells that increasingly extend to periglomerular and tubulointerstitial tissue, as in chronic human GN (Fig. 2c, d) [18]. Deficiency or blockade of proinflammatory chemokines [21] and cytokines [22] reduces renal leukocyte infiltrates and ameliorates glomerular and secondary tubulointerstitial tissue injury, demonstrating the pivotal role of these mediators in the pathogenesis of NTN.

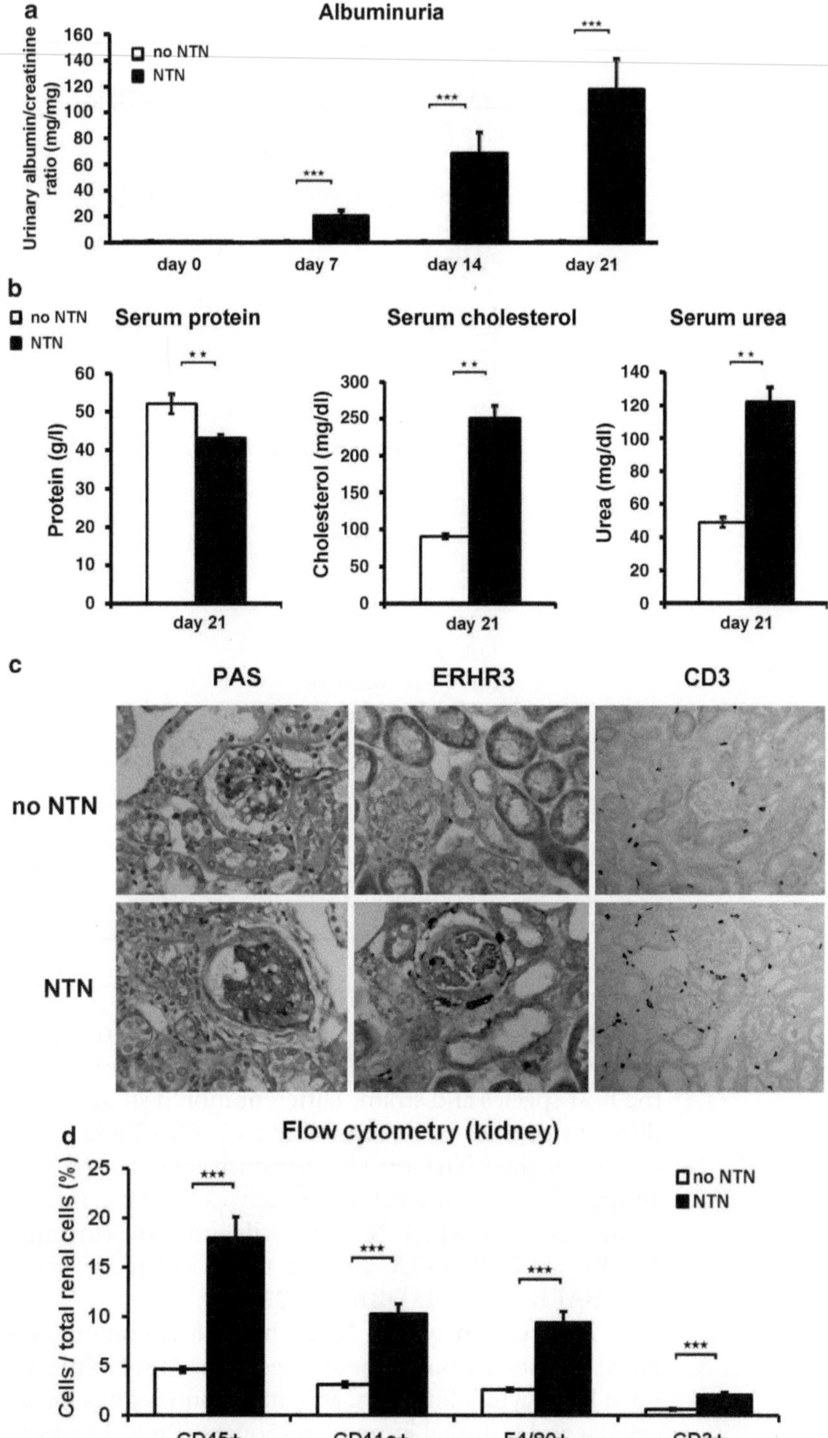

Fig. 2 Phenotype analysis of mice with nephrotoxic serum nephritis (NTN). Data shown are from an accelerated autologous NTN model induced in male wild-type C57BL/6 mice with rabbit nephrotoxic serum. (**a**) Albuminuria expressed as albumin to creatinine ratio. (**b**) Serum parameters demonstrating hypoproteinemia, hypercholesterolemia, and increased blood urea nitrogen levels at day 21 of NTN. (**c**) Glomerular injury and renal leukocyte infiltration at day 21 of NTN revealed by periodic acid Schiff (PAS) stain and immunohistochemistry for ERHR3$^+$ macrophages (original magnification 400×) and CD3$^+$ T cells (original magnification 200×) on paraffin sections. (**d**) Flow-cytometric analysis of renal leukocyte infiltrates at day 21 of autologous NTN. Data are mean ± SEM of 8–14 mice. $^{***}p<0.001$, $^{**}p<0.01$

Table 1
Strain-dependent susceptibility to NTN in mice [23–25]

Inbred mouse strains susceptible to NTN
129/sv
BUB/Bn
C57BL/6
C58
CBA
DBA/1
NZW
Inbred mouse strains not susceptible to NTN
A/J
ALR/Lt
AKR/J
BALB/c
C3H/He
CAST/Ei
DBA/2
DDY/JclSidSeyFrk
FVB/N
MRL/MpJ
NOD/Lt
PERA/Ei
P/J
SB/Le
SJL/J
SWR/J

Inflammatory features of NTN depend strongly on both the donor species, from which the nephrotoxic serum originated, and the host species and strain. Different inbred strains of mice possess different degrees of susceptibility to NTN (Table 1), only in part related to their Th1 or Th2 responsiveness to the nephritogenic antigens [23]. Strains can be broadly categorized into two groups, namely, those in which NTN is easily inducible (including 129/sv, C57BL/6, C58, DBA/1), and those which are not susceptible (like BALB/c, C3H, DBA/2) [23–25].

Phenotype analysis of NTN should focus on functional parameters, extent of glomerular and secondary tubulointerstitial injury, renal leukocyte infiltration, and, in the autologous model, adaptive cellular and humoral immune responses. Table 2 gives an overview on the applied methods and their implications in NTN.

The following protocols describe the preparation of anti-mouse nephrotoxic serum in rabbits, the induction of heterologous and autologous NTN in C57BL/6 mice, possible interventions to stimulate TLR signaling during the NTN course, and the preparation of urine, serum, and tissue samples for subsequent phenotype analysis.

Table 2
Methodology for phenotype analysis in NTN

Method	Parameter	Implication
Functional parameters		
ELISA	Spot urine albumin	Albuminuria
Colorimetric assay (alkaline picric acid method)	Spot urine creatinine	Quantification of albuminuria as albumin to creatinine ratio
Autoanalyzer	Serum creatinine	Renal function
Autoanalyzer	Serum urea	Renal function
Autoanalyzer	Serum protein	Hypoproteinemia
Autoanalyzer	Serum albumin	Hypoalbuminemia
Autoanalyzer	Serum cholesterol	Hypercholesterolemia
Renal injury		
Histology	Periodic acid Schiff (PAS) stain	Glomerulosclerosis score Number of crescents
Histology	Flattened/atrophic tubuli (PAS)	Tubular injury score
Histology	Silver methenamine stain	Glomerular matrix deposition
Immunohistochemistry	Fibrin staining	Glomerular fibrinoid necrosis
Immunofluorescence	Podocyte staining (WT1 and nephrin)	Podocyte loss
Immunofluorescence	TUNEL staining	Necrotic cells in glomeruli and tubulointerstitium
qRT-PCR	mRNA expression	Glomerular (nephrin) and tubular (KIM-1, NGAL) expression of injury markers
qRT-PCR	mRNA expression	Renal expression of inflammatory mediators (chemokines, cytokines, etc.)
ELISA	Chemokine, cytokine concentration in tissue supernatants	Renal expression of inflammatory mediators (chemokines, cytokines)
Immunoblotting	Protein expression	Renal expression of inflammatory mediators (chemokines, cytokines)
Renal leukocyte infiltration		
Flow cytometry	CD45+ Ly6G+renal cells	Renal neutrophil infiltration
Flow cytometry	CD45+ F4/80+ renal cells	Renal mononuclear phagocyte infiltration
Flow cytometry	CD45+ CD11c+cells	Renal dendritic cell infiltration
Flow cytometry	CD45+ CD3+ CD4+ cells	Renal CD4 T cell infiltration
Flow cytometry	CD45+ CD3+ CD8+ cells	Renal CD8 T cell infiltration
Immunohistochemistry	F4/80, ER-HR3 staining	Glomerular and interstitial phagocyte infiltration
Immunohistochemistry	CD3 staining	Glomerular and interstitial T cell infiltration

(continued)

Table 2 (continued)

Method	Parameter	Implication
Local and systemic immune responses		
Immunohistochemistry	Glomerular autologous antibody staining	Local humoral immunity
Immunohistochemistry	Glomerular complement C3d staining	Local humoral immunity
ELISA	Autologous antibody titers (*see* **Note 14**)	Systemic humoral immunity
Intradermal injection of heterologous IgG	Thickness of ear swelling in pre-sensitized animals	Delayed type hypersensitivity response
ELISA	Cytokine concentration (IFN-γ) in splenocyte supernatants after rechallenge with heterologous IgG	Systemic cellular immunity (T cell activation)
Flow cytometry	CD69 surface expression in CD3 + CD4+ and CD3 + CD8+ spleen cells after rechallenge with heterologous IgG	Systemic cellular immunity (T cell activation)

For detailed protocols on the different methodology applied for phenotype analysis (Table 2) the reader is referred to specific literature describing the techniques of ELISA, standard histology on paraffin-embedded tissue sections, immunohistochemistry, immunofluorescence, and flow-cytometric analysis.

2 Materials

2.1 Preparation of Mouse GBM Antigen

1. 500 mL beaker, 50 mL tubes, 1 mL screw-top tubes, non-coated petri dishes, scalpels.
2. 150 μm, 75 μm, 45 μm steel sieves, sterilized (VWR International, Radnor, PA, USA).
3. 1 M NaCl, distilled water, phosphate buffered saline (PBS), all sterile.
4. Sonicator with microtip.

2.2 Immunization of Donor Animal (Rabbit) and Nephrotoxic Serum Preparation

1. Male New Zealand White (NZW) rabbits, specific pathogen-free (SPF), weighing approx. 2.5 kg.
2. Murine GBM suspension (as prepared following the protocol described in Subheading 3.1).
3. Complete Freund's adjuvant (CFA), incomplete Freund's adjuvant (IFA).
4. 0.45 and 0.2 μm syringe filters, low protein binding.

2.3 Immunization of the Host Animal (Mouse)

1. 11 mg/mL Rabbit IgG (ChromPure rabbit IgG, Jackson ImmunoResearch, West Grove, PA, USA).
2. Complete Freund's adjuvant (CFA).
3. Phosphate buffered saline (PBS).
4. 22G needles, 1 mL syringes.
5. Sonicator with microtip.

2.4 Induction of NTN in Mice

1. Aliquots of nephrotoxic serum.
2. Mouse retainer.
3. 30G needles, 1 mL syringes.
4. 70 % ethanol.

2.5 Stimulation of TLR Signaling During NTN

1. Sterile normal saline, dimethylsulfoxide (DMSO).
2. Aliquots of TLR agonists dissolved in normal saline, including endotoxin-free poly(I:C) RNA, ultrapure lipopolysaccharide (LPS), and endotoxin-free unmethylated cytosine-guanine (CpG)-phosphothioate oligodeoxynucleotide 1668 (5′-TCGA TGACGTTCCTGATGCT-3′).
3. Lipofectamine 2000.
4. Pam_3CysSK_4 and imiquimod.
5. 26G needles, 1 mL syringes.
6. 70 % ethanol, swabs.

3 Methods

3.1 Preparation of Mouse GBM Antigen

1. Dissect kidneys from C57BL/6 mice. Remove renal capsule by gently squeezing the kidneys between two fingers. Collect kidneys in 50 mL tubes with PBS and store at −20 °C until needed for further preparation.
2. In a non-coated petri dish, isolate renal cortex by mincing the kidneys crudely. Collect tissue in a separate beaker with PBS on ice.
3. Using a few cortices at a time (so the sieve will not clog) sieve isolate glomeruli using a series of steel sieves of decreasing pore size (150-, 75-, and 45-μm mesh in a top to bottom order). Use PBS for washing (in a wash bottle) through sieves. Collect glomeruli from the 45-μm sieve into new 50 mL tubes, avoiding the transfer of glomerular suspensions with large fluid volumes. Examine the first preparation under the microscope to see that most of the material is glomeruli. Every once in a while collect from the 75-μm sieve. Sieves clog easily, so make sure to rinse between each aliquot.
4. Centrifuge and aspirate supernatant (at 350×*g*, for 10 min at 4 °C).

5. Resuspend the pellet into a 50 mL tube with 30 mL 1 M NaCl. Vortex to break up pellet.
6. Sonicate 4 times for 30 s.
7. Centrifuge at 350 × *g*, 4 °C for 10 min with a low brake.
8. Wash 2 times with 25 mL 1 M NaCl. Centrifuge as in **step 7**.
9. Wash 3 times in 20 mL sterile distilled water. Centrifuge as in **step 7**.
10. Resuspend pellet in distilled water to form approx. 25 % v/v (pellet to water).
11. Aliquot murine GBM antigen suspension into 1 mL screw-top tubes and freeze at −20 °C until donor animal immunization.

3.2 Immunization of Donor Animal (Rabbit) and Nephrotoxic Serum Preparation

Several standard protocols are available for rabbit polyclonal antibody production that are optimized to maximize titer and volume. The protocol described below ensures sufficient time for full maturation of the secondary immune response and serum antibody production in the rabbit. However, protocols may vary according to local practices (*see* **Note 4**).

1. Day 0: Obtain pre-immune bleed (5 mL per rabbit) for control serum collection.
2. Day 1: Immunize rabbit with a primary injection of approx. 500 μL murine GBM antigen suspension mixed with the same volume of CFA. Inject antigen with a 22G needle in multiple (5–10) subcutaneous sites along the back, with 0.1–0.2 mL/injection. A maximum volume of 1.0 mL is injected per rabbit.
3. Day 21: Re-immunize (booster) the rabbit with approx. 500 μL murine GBM antigen suspension mixed with the same volume of IFA. Inject antigen as in **step 1**, with subcutaneous injection sites located on the opposite side of the back from the initial immunizations. Use alternate sites for additional immunizations.
4. Day 42: Perform next subcutaneous booster injection with approx. 250 μL murine GBM antigen suspension mixed with IFA, again at 1:1 ratio with the antigen.
5. Day 52: Draw a first test bleed (approx. 5 mL serum). At this stage an ELISA titer assay (using an aliquot of the GBM antigen suspension for coating ELISA plates) can be performed to evaluate the effect of the antigen on the rabbit.
6. Day 63: Perform next subcutaneous booster injection as in **step 4**.
7. Day 73: Draw production bleed (approx. 20 mL serum).
8. Further re-immunize the rabbit in a 3-week cycle of boosts, with production bleeds (approx. 20 mL serum) taken 10 days after the boosts.

9. Combine serum batches of 4–5 bleeds (approx. 100 mL serum) into one aliquot to be characterized later in vivo for its potency to induce NTN. The serum can be stored at this stage at −20 °C or −80 °C.
10. Heat serum at 56 °C for 30 min in a water bath to inactivate complement, cool down to room temperature for 30 min (*see* **Note 5**).
11. Filter-sterilize the serum by subsequent passage through the 0.45 and 0.20 μm syringe filters, aliquot into sterile 1.5 mL microcentrifuge tubes and store at −20 °C or −80 °C until use (*see* **Note 6**).

3.3 Immunization of the Host Animal (Mouse)

The accelerated autologous model of NTN can be induced by injection of subnephritic doses of nephrotoxic serum into hosts pre-immunized to immunoglobulins of the donor species. Of note, autologous NTN may also develop without pre-immunization after injecting sufficient doses of nephrotoxic serum. When the heterologous model is studied, no additional immunization procedures are performed.

1. Mix 36 μL ChromPure rabbit IgG (11 mg/mL) with 464 μL sterile PBS in a 50 mL tube, add 1.5 mL PBS resulting in a 0.2 mg/mL rabbit IgG solution.
2. Add 2 mL complete Freund's adjuvant to 2 mL rabbit IgG solution (0.2 mg/mL), vortex and store on ice.
3. Prepare into an emulsion by sonicating the solution with high power on ice for 30 s, cool for 30 s on ice and repeat the procedure 5–6 times until emulsion is thick and creamy (*see* **Note 7**).
4. 3 days before administration of the nephrotoxic serum, inject 100 μL of the emulsion subcutaneously in each flank of the mouse using a 26G needle (20 μg donor IgG per mouse).

3.4 Induction of NTN in Mice and Sample Preparation

As noted above, the dose of nephrotoxic serum to induce heterologous, autologous or accelerated autologous NTN will vary, depending on the donor and host strains, antigenicity of the GBM preparations, efficacy of the host's immunization during nephrotoxic serum generation, host strain, and also host gender (*see* **Note 8**). Different serum aliquots obtained from the same donor may also perform differently. In addition, the time course of renal functional alterations and the extent of renal injury depend on the abovementioned factors. Therefore, the applied dose of the nephrotoxic serum must be determined empirically, depending on the specific experimental setting. Here, we describe induction of heterologous and accelerated autologous NTN with rabbit nephrotoxic serum in male C57/BL6 mice.

1. Collect spot urine samples for day 0 values before serum injection (*see* **Note 9**).
2. Fixate 7–9-week-old male mouse in retainer.
3. Inject 400 μL (heterologous model) or 100 μL (autologous modes) of rabbit nephrotoxic serum intravenously into the tail vein using a 30G needle after cleaning the injection site with 70 % ethanol.
4. Heterologous model: Collect spot urine samples (preferably at least 50–100 μL) in regular intervals for analysis of albuminuria, e.g., at 2 h, day 2, 5, and 7 (*see* **Notes 10** and **11**). The experiment may be terminated at any of these time points to obtain serum and renal tissue for subsequent phenotype analysis (see Table 2).
5. Autologous model: Collect spot urine samples (preferably at least 50–100 μL) and terminate experiment at day 7, 14, or 21, depending of the phenotypic features of the experimental setting (*see* **Notes 10** and **11**).
6. At termination of the experiment anesthetize the mouse according to local standards. Obtain blood samples for serum analysis by retroorbital bleeding. Fixate the mouse, perform a midline abdominal incision, and harvest both kidneys and spleen as required (*see* **Note 12**).
7. For renal phenotype analysis (Table 2) we routinely collect the following specimens:
   (a) Upper (optional) and lower third of right kidney for preparation of single cell suspensions and subsequent flow cytometry.
   (b) Middle portion of right and left kidney for fixation in formalin, paraffin embedding, and subsequent routine histology and immunohistochemistry.
   (c) Upper third of left kidney for mRNA expression analysis.
   (d) Lower third of left kidney for placing in OTC freezing compound and cryostat sectioning if required.
   (e) Alternatively, lower third of one or both kidneys can be used for protein analysis with immunoblotting.

3.5 Stimulation of TLR Signaling During NTN

Synergistic stimulation of TLRs during the course of NTN can be achieved by intraperitoneal injection of TLR agonists, including Pam_3CysSK_4 (for TLR2 activation), poly(I:C) RNA (TLR3), LPS (TLR4), imiquimod (TLR7), and CpG oligodeoxynucleotides (TLR9). TLR agonists are injected intraperitoneally on alternate days during NTN from day 0–21 (or other time periods as required, *see* **Note 13**).

1. Dissolve required amounts of TLR agonists in an appropriate volume of sterile normal saline or sterile endotoxin-free water. The following doses are used for each injection and mouse: 10 μg of LPS or 40 μg of CpG oligonucleotides dissolved in 100 μL of normal saline, or 50 μg of poly(I:C) RNA dissolved in 100 μL of sterile endotoxin-free water. 100 μL of normal saline or sterile endotoxin-free water serves as vehicle control.
2. To enhance intracellular uptake of poly(I:C) RNA in vivo, it can be complexed to cationic lipids before injection (poly(I:C) RNA–cationic lipid complexes). These complexes are generated by incubation of 50 μL of poly(I:C) RNA previously dissolved in sterile endotoxin-free water with 50 μL of the cationic lipid Lipofectamine 2000 (1:1, vol:vol). 50 μL of the cationic lipid Lipofectamine 2000 mixed with 50 μL of sterile endotoxin-free water serves as vehicle control.
3. Pam_3CysSK_4 and imiquimod are dissolved in sterile endotoxin-free water. For each intraperitoneal injection 100 μg of Pam_3CysSK_4 or 25 μg of imiquimod in 100 μL of sterile endotoxin-free water is used. 100 μL of sterile endotoxin-free water serves as vehicle control.
4. Before intraperitoneal injection clean abdominal injection sites with swabs soaked in 70 % ethanol. Every second day each mouse is injected with either 100 μL of the vehicle solution or 100 μL of vehicle containing the respective TLR agonist.

4 Notes

1. We prefer not to use the term anti-GBM-glomerulonephritis for NTN. The nephrotoxic serum not only contains antibodies directed against GBM constituents like type 4 collagen but may also show, for example, reactivity with several podocyte cell surface antigens [26]. In contrast, experimental autoimmune anti-GBM glomerulonephritis is a different model that can be induced by actively immunizing mice or rats with the non-collagenous domain of the α3 chain of type 4 collagen, which results in a T helper cell-dependent production of autoantibodies against GBM [27].
2. Nephrotoxic serum generated by immunization of host strains with GBM antigen prepared from rat (and other species) induces NTN in mice [28]. Such nephrotoxic sera are available from commercial vendors, which induce a robust NTN phenotype.
3. Nephrotoxic serum produced in both rabbits and sheep are suitable for NTN induction in mice and are widely used.
4. We recommend that a minimum of two rabbits is used for generation of nephrotoxic serum to encounter for biological variations.

5. We use heat-inactivated whole nephrotoxic serum. Alternatively, a globulin fraction can be obtained by precipitation in 50 % ammonium sulfate, resuspension in PBS and subsequent dialysis against PBS. Other investigators purify antibody subclasses through absorption procedures.
6. Aliquots of nephrotoxic serum should be checked for endotoxin content before use, as lipopolysaccharide (LPS)-induced TLR activation will exacerbate NTN [29]. LPS content can be measured, for example, with the 1000-QCL LPS assay kit from Biowhittaker (Walkersville, MD, USA).
7. To test the quality of the antigen emulsion used for immunization of mice in the accelerated autologous model aspirate 200 μL of the emulsion after sonication and drop into water. The emulsion should form a drop for about 10 s before dispensing.
8. NTN develops differently in male and female mice, so we only use male mice aged 7–9 weeks for experiments.
9. The baseline urine sample must be obtained before first injection of nephrotoxic serum, as proteinuria may commence shortly after the injection.
10. We do not place mice in metabolic cages for urine collection to determine urinary 24 h albumin excretion rates. This is stressful to the mice and can be avoided by collecting spot urine samples. ELISA-determined albumin concentrations in spot urine are normalized to creatinine concentrations in the same sample analyzed by a colorimetric assay (alkaline picric acid method) as an indicator for urine concentration. Albuminuria is expressed as urinary albumin to creatinine ratio (ACR). 10 μL of urine is usually a sufficient volume to determine albumin and creatinine concentrations.
11. Proteinuria semiquantitatively measured by commercially available dipsticks is a good screening parameter to test the potency of different nephrotoxic serum aliquots to induce NTN.
12. Although mice develop a nephrotic syndrome-like hypoproteinemia and hypercholesterolemia in the autologous NTN model, we seldom observe overt edema or ascitis in severely proteinuric mice.
13. TLR agonists may also be administered only once at the time of nephrotoxic serum injection (heterologous model) [15, 16] or at immunization (autologous model) [30] to analyze influences on innate immunity during the induction phase of NTN or the adaptive immune response, respectively.
14. Proteinuric mice lose substantial amounts of IgG in the urine, leading to low serum IgG levels. This also affects autologous antibody titers. Mice which are protected from disease (i.e., less

proteinuric) will often have higher titers than diseased animals, which does not necessarily indicate differences in the systemic humoral immune response.

Acknowledgements

V.V. was supported by DFG grants VI 231/2-1 and VI 231/2-2.

References

1. Unanue ER, Dixon FJ (1967) Experimental glomerulonephritis: immunologic events and pathogenic mechanisms. Adv Immunol 6:1–90
2. Dixon FJ, Wilson CB (1990) The development of immunopathologic investigation of kidney disease. Am J Kidney Dis 16:574–578
3. Smadel JE (1937) Experimental nephritis in rats induced by injection of anti-kidney serum: III. pathological studies of the acute and chronic stages. J Exp Med 65:541–555
4. Smadel JE, Farr LE (1937) Experimental nephritis in rats induced by injection of anti-kidney serum: II. clinical and functional studies. J Exp Med 65:527–540
5. Unanue ER, Dixon FJ (1965) Experimental glomerulonephritis. VI. The autologous phase of nephrotoxic serum nephritis. J Exp Med 121:715–725
6. Shirato I, Sakai T, Kimura K et al (1996) Cytoskeletal changes in podocytes associated with foot process effacement in Masugi nephritis. Am J Pathol 148:1283–1296
7. Hammer DK, Dixon FJ (1963) Experimental glomerulonephritis. II. Immunologic events in the pathogenesis of nephrotoxic serum nephritis in the rat. J Exp Med 117:1019–1034
8. Sheerin NS, Springall T, Carroll MC et al (1997) Protection against anti-glomerular basement membrane (GBM)-mediated nephritis in C3- and C4-deficient mice. Clin Exp Immunol 110:403–409
9. Hébert MJ, Takano T, Papayianni A et al (1998) Acute nephrotoxic serum nephritis in complement knockout mice: relative roles of the classical and alternative pathways in neutrophil recruitment and proteinuria. Nephrol Dial Transplant 13:2799–2803
10. Schrijver G, Bogman MJ, Assmann KJ et al (1990) Anti-GBM nephritis in the mouse: role of granulocytes in the heterologous phase. Kidney Int 38:86–95
11. Quigg RJ (2004) Complement and autoimmune glomerular diseases. Curr Dir Autoimmun 7:165–180
12. Lin F, Salant DJ, Meyerson H et al (2004) Respective roles of decay-accelerating factor and CD59 in circumventing glomerular injury in acute nephrotoxic serum nephritis. J Immunol 172:2636–2642
13. Couser WG, Stilmant MM, Jermanovich NB (1977) Complement-independent nephrotoxic nephritis in the guinea pig. Kidney Int 11: 170–180
14. Quigg RJ, Kozono Y, Berthiaume D et al (1998) Blockade of antibody-induced glomerulonephritis with Crry-Ig, a soluble murine complement inhibitor. J Immunol 160:4553–4560
15. Fu Y, Xie C, Chen J et al (2006) Innate stimuli accentuate end-organ damage by nephrotoxic antibodies via Fc receptor and TLR stimulation and IL-1/TNF-α production. J Immunol 176: 632–639
16. Brown HJ, Lock HR, Sacks SH et al (2006) TLR2 stimulation of intrinsic renal cells in the induction of immune-mediated glomerulonephritis. J Immunol 177:1925–1931
17. Flür K, Allam R, Zecher D et al (2009) Viral RNA induces type I interferon-dependent cytokine release and cell death in mesangial cells via melanoma-differentiation-associated gene-5: implications for viral infection-associated glomerulonephritis. Am J Pathol 175:2014–2022
18. Tipping PG, Holdsworth SR (2006) T cells in crescentic glomerulonephritis. J Am Soc Nephrol 17:1253–1263
19. Paust HJ, Turner JE, Steinmetz OM et al (2009) The IL-23/Th17 axis contributes to renal injury in experimental glomerulonephritis. J Am Soc Nephrol 20:969–979
20. Hochheiser K, Engel DR, Hammerich L et al (2011) Kidney dendritic cells become pathogenic during crescentic glomerulonephritis with proteinuria. J Am Soc Nephrol 22: 306–316
21. Vielhauer V, Anders HJ (2009) Chemokines and chemokine receptors as therapeutic targets in chronic kidney disease. Front Biosci (Schol Ed) 1:1–12

22. Tipping PG, Holdsworth SR (2007) Cytokines in glomerulonephritis. Semin Nephrol 27: 275–285
23. Huang XR, Tipping PG, Shuo L et al (1997) Th1 responsiveness to nephritogenic antigens determines susceptibility to crescentic glomerulonephritis in mice. Kidney Int 51:94–103
24. Xie C, Sharma R, Wang H et al (2004) Strain distribution pattern of susceptibility to immune-mediated nephritis. J Immunol 172:5047–5055
25. Xie C, Rahman ZS, Xie S et al (2008) Strain distribution pattern of immune nephritis – a follow-up study. Int Immunol 20:719–728
26. Chugh S, Yuan H, Topham PS et al (2001) Aminopeptidase A: a nephritogenic target antigen of nephrotoxic serum. Kidney Int 59: 601–613
27. Kitching AR, Turner AL, Semple T et al (2004) Experimental autoimmune anti-glomerular basement membrane glomerulonephritis: a protective role for IFN-γ. J Am Soc Nephrol 15:2373–2382
28. Nishihara T, Kusuyama Y, Gen E et al (1981) Masugi nephritis produced by the antiserum to heterologous glomerular basement membrane. I. Results in mice. Acta Pathol Jpn 31:85–92
29. Karkar AM, Rees AJ (1997) Influence of endotoxin contamination on anti-GBM antibody induced glomerular injury in rats. Kidney Int 52:1579–1583
30. Brown JH, Sacks SH, Robson MG (2006) Toll-like receptor 2 agonists exacerbate accelerated nephrotoxic nephritis. J Am Soc Nephrol 17:1931–1939

Chapter 16

Isolation of Intratumoral Leukocytes of TLR-Stimulated Tumor-Bearing Mice

Moritz Rapp, David Anz, and Max Schnurr

Abstract

Toll-like receptor (TLR) ligands hold promise for cancer immunotherapy. The isolation of intratumoral leukocytes of tumor-bearing mice is a useful technique for analyzing the immunological effects of TLR ligands on the tumor microenvironment. These isolated immune cells can be directly used for analysis (e.g., by flow cytometry) or cultured for functional in vitro studies. Here, we describe the isolation of intratumoral leukocytes by density gradient centrifugation. This technique can be used to isolate leukocytes from freshly dissected murine tumors.

Key words Leukocyte isolation, TLR activation, Density gradient centrifugation, Murine tumor models

1 Introduction

TLR ligands activate the innate immune system and can be exploited to break tumor-induced immunosuppression [1]. Treating mice with experimental tumors with CpG oligonucleotides, synthetic TLR9 ligands, can reduce tumor growth and increase overall survival [2–4]. Promising data from animal studies has prompted clinical trials investigating the anti-tumoral effects of TLR9 ligands in humans [5, 6]. The synthetic TLR7 agonist imiquimod is another example for effective TLR-mediated anticancer therapy. This TLR7 ligand is FDA approved and used in the clinic to treat superficial basal cell carcinoma and vulvar intraepithelial neoplasia [7, 8]. Nevertheless, the precise mechanisms of TLR-based immunotherapy are still poorly understood. One important approach to investigate these mechanisms in vivo is to isolate tumor-infiltrating immune cells from murine tumors after treatment with TLR ligands. In this chapter we describe in detail how to isolate tumor-infiltrating immune cells from murine tumors for subsequent analysis.

Hans-Joachim Anders and Adriana Migliorini (eds.), *Innate DNA and RNA Recognition*, Methods in Molecular Biology, vol. 1169, DOI 10.1007/978-1-4939-0882-0_16, © Springer Science+Business Media New York 2014

2 Materials

1. Phosphate buffered saline (PBS).
2. Porcine trypsin (2.5 g/l) with 1 mM EDTA.
3. Isoflurane.
4. Scissors and scalpel.
5. Collagenase type I.
6. DNase I type IV.
7. Cell culture medium (e.g., RPMI).
8. Cell strainers, 100 and 40 μm pore size.
9. 44 % Percoll (stock density = 1.124 g/ml) solution (mix 4.4 ml of Percoll and 5.6 ml of PBS).
10. 67 % Percoll (stock density = 1.124 g/ml) solution (mix 6.7 ml of Percoll and 3.3 ml of PBS).

3 Methods

3.1 Tumor Cell Culture and Subcutaneous Tumor Induction

1. One day before tumor induction, split cultured tumor cells in an appropriate ratio (e.g., 1:1) to ensure optimal exponential cell growth (*see* **Note 1**).
2. On the next day discard tumor cell culture medium and remove non-adherent tumor cells by washing the flask with PBS (if non-adherent tumor cells are used, skip this step and proceed with **step 4**).
3. Detach adherent cells by adding trypsin/EDTA to the cells (*see* **Note 2**). After 2–3 min incubation at 37 °C remove all the cells by adding PBS and wash cells by centrifugation at 400 × *g* and 4 °C for 7 min.
4. Repeat washing at least two times to remove all traces of culture medium (*see* **Note 3**).
5. After the last washing step re-suspend cells in a small volume of PBS for counting. Subsequently, adjust the volume to achieve the desired tumor cell density for injection (*see* **Note 4**). For subcutaneous injection do not use more than 200 μl injection volumes per mouse. The most frequent site for subcutaneous tumor injection is the skin over the flank.

3.2 Tumor Size Measurement and Treatment with TLR Ligands

1. Approximately 5–7 days after tumor induction a small mass should be palpable (*see* **Note 5**). Once tumors are palpable start with tumor size measurement. Tumor diameters are measured with a caliper to calculate tumor surface (width × length).
2. When the tumors reach an average size of 25–65 mm^2 treatment with TLR ligands should be started (*see* **Note 6**).

Most TLR ligands are injected subcutaneously into the flank opposite of the tumor. Alternative routes are intraperitoneal or intravenous injections. Usually, the treatment is repeated at least twice a week to achieve effects on tumor growth and tumor immune cell infiltration.

3.3 Tumor Resection

1. Depending on the proliferation rate of the tumor cell line and the initial tumor cell number used for tumor induction tumors will be explanted 3–6 weeks after tumor induction. Usually, mice are anesthetized with isoflurane and sacrificed by cervical dislocation when the tumor reaches a size of approximately 150–200 mm^2 to obtain sufficient tissue for subsequent analysis. The complete tumor is surgically removed (*see* **Note 7**).
2. Immediately homogenize the tumors by using scissors and a scalpel and incubate the obtained tumor homogenates for 30 min at 37 °C on a shaker with 1 mg/ml collagenase type I and 0.05 mg/ml DNase I type IV in cell culture medium (*see* **Note 8**).
3. To prepare a single cell suspension, the homogenate is meshed through a 100 μm cell strainer followed by a 40 μm cell strainer.
4. Finally pellet the cells by centrifugation and re-suspend the pellet in 2–5 ml PBS.

3.4 Isolation of Intratumoral Leukocytes by Density Gradient Centrifugation

1. For gradient centrifugation two different Percoll layers are needed. Prepare a 44 % and a 67 % Percoll solution by mixing the appropriate amount of Percoll ($d = 1.124$ g/ml) with PBS.
2. Place 9 ml of the 44 % Percoll solution into a 50 ml centrifugation tube. Underlay the 44 % solution with 6 ml of the 67 % solution (Fig. 1a). To avoid mixing of the two layers pipetting must be done very slowly and gently (*see* **Note 9**).
3. Finally, layer the re-suspended single cell suspension obtained from the tumor tissue on top of the upper layer and perform centrifugation at 800 × *g* at 4 °C for 30 min with brake enabled (*see* **Note 10**).
4. After centrifugation a white ring of leukocytes is visible at the interphase of the two Percoll layers. Use a pipette to obtain the leukocytes of this ring (Fig. 1b).
5. Finally, wash the obtained leukocytes once with PBS. The isolated cells can be analyzed directly (e.g., by flow cytometry) or can be cultured in vitro for studying their functional properties.

4 Notes

1. For optimal tumor growth in vivo it is mandatory to use tumor cells that are in their exponential growth phase. Therefore, splitting tumor cells 1 day before injection is important.

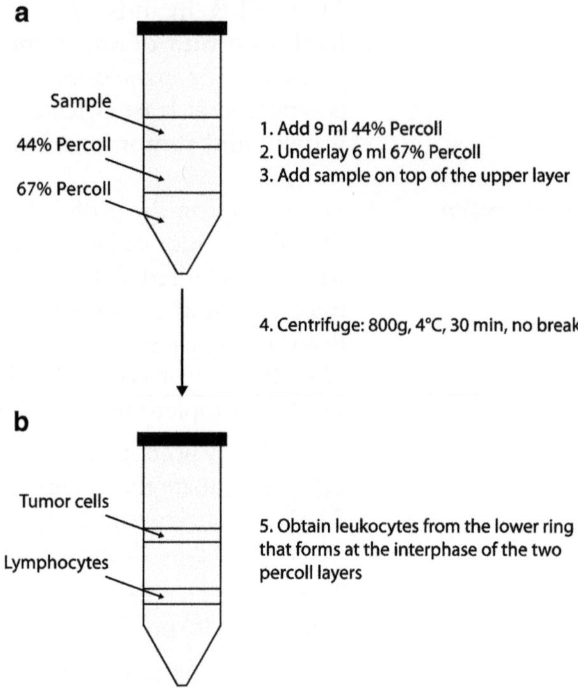

Fig. 1 Pipetting schema for Percoll density gradient centrifugation

2. For a 175 cm^2 cell culture flask use 1 ml 1× trypsin/EDTA. For smaller flasks use accordingly less volume of trypsin/EDTA.
3. Extensive washing minimizes unintended immune reactions and irritation in mice due to residual immunogenic substances in the cell culture medium. Thorough washing may be obligatory to prevent tumor rejection.
4. Typically mice are injected with 2×10^5 to 1×10^6 tumor cells. For slow-growing tumor cell lines it is advisable to inject more tumor cells.
5. For convenient monitoring of tumor growth, the skin should be shaved with a small razor in the area of tumor cell injection.
6. Generally, larger tumors are more difficult to treat as compared to small tumors, as tumor size correlates with immunosuppression. Thus, the time point of therapy commencement should be adapted to the expected potency of the treatment.
7. A thin skin layer covers subcutaneous tumors. Use scissors and tweezers to detach the tumor from this skin layer. Also remove blood vessels that are incorporated into the tumor due to angiogenesis.

8. Collagenase and DNase treatment is mandatory for optimal immune cell isolation from the tumor cell suspension. For tumors with an average size of 150–200 mm^2, 2 ml cell culture medium (with 1 mg/ml collagenase type I and 0.05 mg/ml DNase I type IV) should be used.
9. To underlay the 44 % Percoll layer with the 67 % Percoll solution use a pipetus and set it to atmospheric pressure emptying.
10. Turning off the brake of the centrifuge is important to avoid disruption of the density gradient.

References

1. Kanzler H, Barrat FJ, Hessel EM, Coffman RL (2007) Therapeutic targeting of innate immunity with Toll-like receptor agonists and antagonists. Nat Med 13(5):552–559
2. Heckelsmiller K, Rall K, Beck S, Schlamp A, Seiderer J, JahrsdoÅNrfer B, Krug A, Rothenfusser S, Endres S, Hartmann G (2002) Peritumoral CpG DNA elicits a coordinated response of CD8 T cells and innate effectors to cure established tumors in a murine colon carcinoma model. J Immunol 169(7):3892–3899
3. Houot R, Levy R (2009) T-cell modulation combined with intratumoral CpG cures lymphoma in a mouse model without the need for chemotherapy. Blood 113(15):3546–3552
4. Krieg AM (2008) Toll-like receptor 9 (TLR9) agonists in the treatment of cancer. Oncogene 27:161–167
5. Weber JS, Zarour H, Redman B, Trefzer U, O'Day S, van den Eertwegh AJ, Marshall E, Wagner S (2009) Randomized phase 2/3 trial of CpG oligodeoxynucleotide PF-3512676 alone or with dacarbazine for patients with unresectable stage III and IV melanoma. Cancer 115(17):3944–3954
6. Vollmer J, Krieg AM (2009) Immunotherapeutic applications of CpG oligodeoxynucleotide TLR9 agonists. Adv Drug Deliv Rev 61:195–204
7. Chang YC, Madkan V, Cook-Norris R, Sra K, Tyring S (2005) Current and potential uses of imiquimod. South Med J 98(9):914–920
8. van Seters M, van Beurden M, ten Kate FJ, Beckmann I, Ewing PC, Eijkemans MJ, Kagie MJ, Meijer CJ, Aaronson NK, Kleinjan A, Heijmans-Antonissen C, Zijlstra FJ, Burger MP, Helmerhorst TJ (2008) Treatment of vulvar intraepithelial neoplasia with topical imiquimod. N Engl J Med 358(14):1465–1473

Chapter 17

Bifunctional siRNAs for Tumor Therapy

Fanny Matheis and Robert Besch

Abstract

Double-stranded RNA molecules carrying a triphosphate moiety represent a molecular structure by which the host recognizes viral infections. Such RNA molecules can be generated synthetically by chemical synthesis or by in vitro transcription (*see* Chapter 2, Hornung et al.). Similar to viruses, they initiate an antiviral immune response, e.g., by stimulation of the immune system. Short, double-stranded RNA in the cytosol can also trigger the RNA interference mechanism, which also has been considered as an antiviral response. Notably, synthetic RNAs that are designed to be specific for a certain host mRNA inhibit expression of the respective gene, leading to specific gene silencing. Both effects—gene silencing and immunostimulation—are interesting from a therapeutic perspective, e.g., for cancer therapy. Notably, both effects can be activated by a single molecule, an siRNA carrying a triphosphate moiety. This chapter provides information how to design such compounds with respect to the associated signaling pathways and the techniques to evaluate bifunctional RNAs in the context of tumor therapy.

Key words 5′Triphosphate-conjugated siRNA, Immunostimulatory RNA, RNA interference, Apoptosis, RIG-like helicases, Interferon induction

1 Introduction

Due to their simple structure, viruses provide a limited amount of molecular patterns that enable the host the detection of a viral infection. Important pathogen-associated molecular patterns (PAMP) for viruses by which the host senses an infection are viral nucleic acids. For viral RNAs, specific pattern recognition receptors (PRR) exist. These include members of the Toll-like receptor family, which detect viral RNA in endosomes and lysosomes and RIG-like helicases (RLH), comprising RIG-I, MDA-5, and LGP-2, which sense viral RNA in the cytosol of infected cells [1]. Short, double-stranded RNAs that carry an uncapped triphosphate residue at their 5′ end (pppRNAs) are detected by the cytosolic receptor RIG-I. Activation of RIG-I enables its translocation and binding to the adapter protein IFN-β promoter stimulator 1 (IPS-1, also called MAVS, Cardif, and VISA) at the mitochondrial membrane.

Hans-Joachim Anders and Adriana Migliorini (eds.), *Innate DNA and RNA Recognition*, Methods in Molecular Biology, vol. 1169, DOI 10.1007/978-1-4939-0882-0_17, © Springer Science+Business Media New York 2014

This initiates a signaling cascade involving interferon-regulatory factors (IRF3 and 7) and NFκB, thereby activating innate immune responses like secretion of type-I interferons and proinflammatory cytokines. Moreover, RIG-I activation initiates a proapoptotic signaling cascade that facilitates apoptosis in infected cells, which is independent of type I interferon signaling.

Another cellular antiviral mechanism initiated by short, double-stranded RNA molecules is RNA interference [2]. Here, double-stranded mRNA in the cytosol (e.g., viral RNA) is processed into short fragments (small interfering RNAs; siRNAs) that are integrated in an enzyme complex called RNA-induced silencing complex (RISC). Both strands are unwound leading to presentation of one of the RNA strands in the cytosol. Association of the RISC complex with the complementary viral RNA via antisense binding triggers cleavage and subsequent degradation of viral mRNA. Synthetic RNAs with the appropriate structure are also integrated in the RISC complex. Notably, they can be designed to be complementary to a certain host mRNA. This directs the antiviral response to host proteins, thereby enabling specific gene silencing via RNA interference [3].

Both mechanisms—immunostimulation and RNA interference—can be activated by short synthetic RNA derivatives. Triphosphate-conjugated RNAs (pppRNAs) can be generated by in vitro transcription or by chemical synthesis. Moreover, both mechanisms can be initiated by one single molecule due to the following characteristics: (1) Both mechanisms are activated in the same cellular compartment, i.e., the cytosol, (2) the RNA sequence plays a minor role for RIG-I activation allowing to design the sequence to target a certain gene of interest, and (3) the triphosphate moiety at the 5′ end does not abrogate RISC integration and activation of RNA interference. The mechanisms that are activated by such bifunctional RNAs (ppp-siRNAs) are illustrated in Fig. 1.

The combination of immune activation and gene silencing is interesting for therapeutic purposes, especially for cancer therapy. The activation of innate immune reactions via RIG-I may help to direct the immune system against the tumor. Moreover, as RIG-I signaling is not restricted to immune cells and present in most cell types, including tumor cells, local expression of interferons and cytokines in tumor and stroma cells may help to circumvent the local tumor immune barrier that prevents immune recognition of the tumor. Furthermore, proapoptotic signaling via RIG-I needs counterregulatory mechanisms to avoid apoptosis. These counterregulatory mechanisms are often defective in cancer cells, leading to apoptosis induction specifically in tumor cells [4, 5]. These effects can be combined with silencing of a therapeutically relevant gene via RNA interference, which allows for tailoring of these compounds for a certain tumor type. The effects elicited by such bifunctional ppp-siRNAs are summarized in Fig. 2.

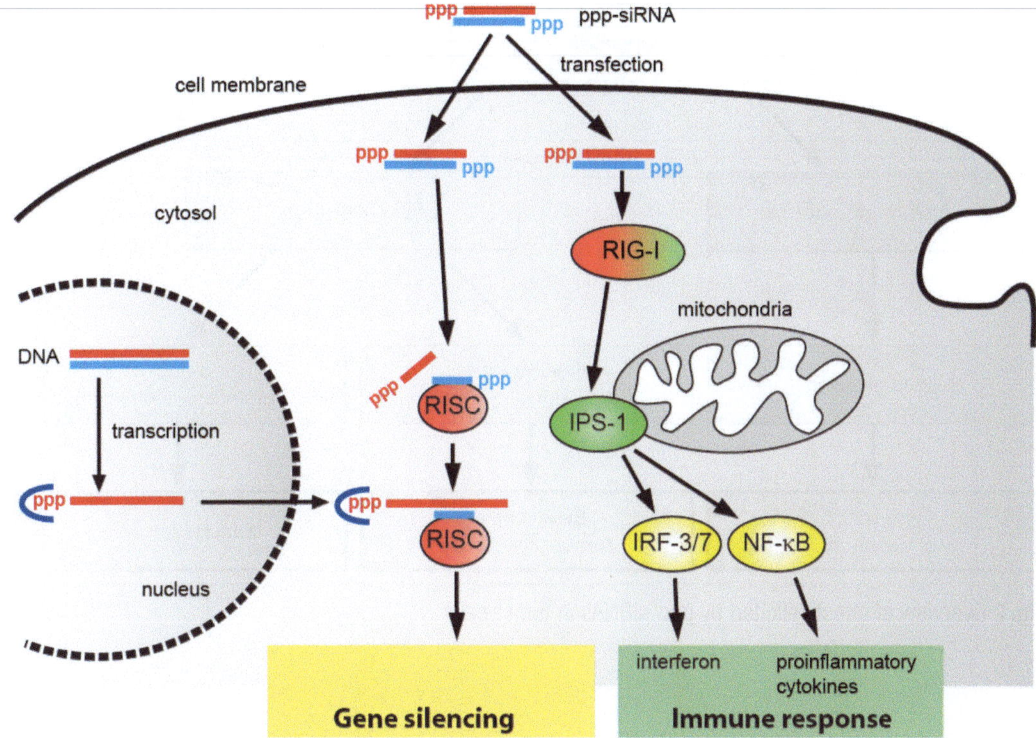

Fig. 1 Cellular mechanisms triggered by bifunctional siRNAs (ppp-siRNA). In the cytoplasm, a ppp-siRNA is integrated into the RISC complex allowing binding of complementary mRNA by antisense recognition. This leads to the degradation of the mRNA and consequently to silencing of the corresponding protein. On the other hand, due to its triphosphate moiety, a ppp-siRNA activates RIG-I, resulting in an antiviral response. RNA generated in the nucleus of eukaryotic cells is single-stranded but carries a triphosphate moiety. However, due to posttranscriptional processing (5′-capping, represented by *dark blue* semicircle) RIG-I activation is avoided

2 Signaling Pathways Associated with Bifunctional RNA Molecules

The mode of action of bifunctional RNAs can be modified by selecting an appropriate gene for RNA silencing. For this, it is important to know the specific signaling pathways that are associated with the RIG-I-activating component of these substances.

2.1 RIG-I-Mediated Immunostimulation

Short, double-stranded pppRNA in the cytosol activates RIG-I [6]. This leads to activation of NFκB and to type-I interferon secretion via the mitochondrial adaptor molecule IPS-1. When selecting a target for gene silencing, it has to be taken into account that a plethora of genes is regulated by NFκB and type-I interferons. Silencing of the gene has to be efficient in this context.

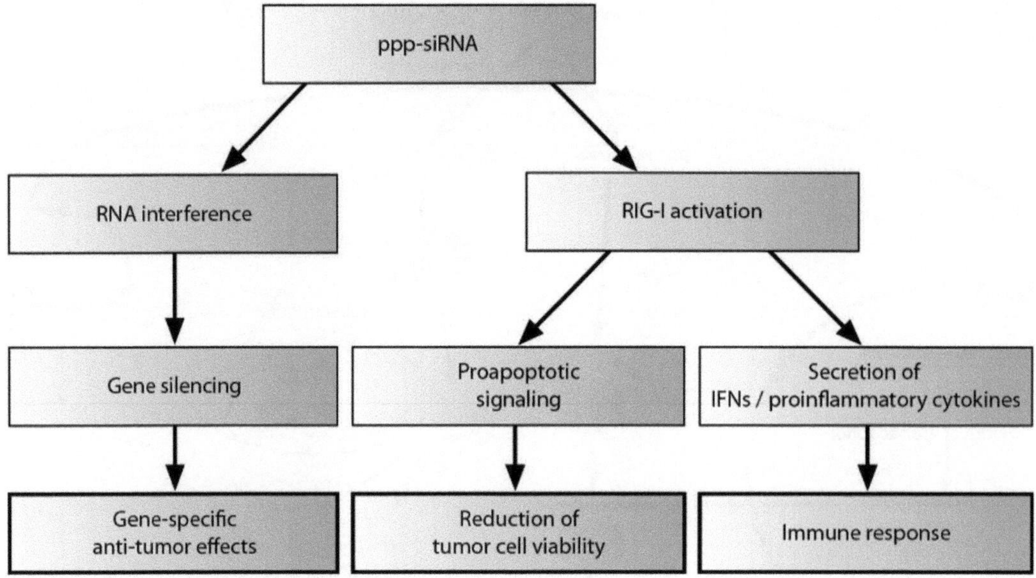

Fig. 2 Overview of effects elicited by ppp-siRNAs in tumor cells

2.2 RIG-I-Mediated Proapoptotic Signaling

In addition to immunostimulation, RIG-I activation initiates proapoptotic signaling consisting of induction of the proapoptotic BH3-only proteins Noxa and Puma. Interestingly, this upregulation was found to be dependent on the adaptor IPS-1, but independent of type I interferon signaling. In primary cells, activation of this signaling pathway does not lead to apoptosis, which presumably is due to the protective function of Bcl-xL, an antiapoptotic protein that counteracts the proapoptotic BH3-only proteins. Apoptosis in infected cells can be interpreted as an antiviral response, however, unbalanced induction of apoptosis of infected cells is likely deleterious for the host. Therefore, counterregulatory mechanisms such as antiapoptotic Bcl-xL signaling are needed to balance this response. Notably, these balancing effects seem to be defective in several types of tumor cells, leading to effective apoptosis induction upon treatment with pppRNAs. This apoptotic effect significantly contributes to the anti-tumor efficacy of bifunctional RNAs in vivo. One characteristic of the RIG-I-triggered apoptotic pathway is that it seems to be independent of the tumor suppressor protein p53. Silencing of p53 did not reduce pppRNA-mediated apoptosis of melanoma cell lines. Notably, Noxa, the key molecule in RIG-I-dependent apoptosis and a well-known p53 target, was activated in a p53-independent way. Nevertheless, induction of Puma required presence of p53; however, Puma was dispensable for apoptosis induction [4].

3 Methods

3.1 Design and Generation of Bifunctional siRNAs

Below, some examples are given for genes whose inhibition is attractive in the context of RIG-I activation.

3.1.1 Target Gene Selection

Genes Whose Inhibition Is Associated with p53 Activation

p53 activation cooperates with apoptotic RIG-I signaling, which is p53-independent. In tumors, in which p53 is not mutated, p53 activity adds to RIG-I signaling not only with respect to apoptosis, but also with respect to the other known tumor-suppressive effects of p53.

Genes Whose Inhibition Activates Complementary Pathways of Cell Death

RIG-I activation initiates the mitochondrial apoptotic pathway. Genes that lead to activation of alternative pathways of cell death, e.g., death receptor-mediated apoptosis or necroptosis, may cooperate with RIG-I signaling.

Inhibition of Oncogenes

Functionally different approaches are of interest for combination with RIG-I signaling. These include oncogenes of high relevance for the tumor type of interest, like proliferation drivers or genes associated with metastasis or angiogenesis.

Genes Associated with Tumor Immune Escape Pathways

Inhibition of immunosuppressive genes may potentiate the immunostimulatory effects of RIG-I. Efficacy of such bifunctional RNAs has been shown for TGF-β1 [7].

3.1.2 siRNA Design

Gene-specific siRNAs can be designed according to published guidelines and several Web-based design algorithms are available to facilitate siRNA design [8–10].

3.1.3 Generation of Triphosphate-Conjugated RNA Molecules

Mostly, ppp-siRNAs are generated via in vitro transcription (IVT), which is described in Chapter 2. Chemical synthesis of pppRNAs has been described; however, due to the complex synthesis, IVT is the most widely used technique to generate ppp-siRNAs.

3.1.4 Controls

RNA Control Setup

Choosing an appropriate setup of controls is crucial. Specifically, it is necessary to include RNA molecules that activate only one of the pathways triggered by bifunctional siRNAs. In addition to the bifunctional ppp-siRNA, (1) a synthetic siRNA counterpart without 5′triphosphate moiety with identical sequence, (2) a non-targeting control-siRNA designed to contain a random sequence not present in the human (or murine, respectively) genome, and (3) a non-targeting control ppp-siRNA, e.g., a in vitro-transcribed triphosphate-conjugated counterpart to the control-siRNA are required. Other controls include a second gene-specific setup with another siRNA sequence to control for sequence-specific effects not related to RNA interference.

Control for Tumor Specificity

The results of proapoptotic signaling differ between non-malignant cells and tumor cells, with apoptosis primarily induced in tumor cells. The comparison of ppp-siRNA effects in these different cell types is of high relevance for therapeutic purposes.

3.2 Evaluation of Bifunctional RNAs

3.2.1 Application of Bifunctional RNAs

Binding of ppp-siRNAs takes place in the cytosol. This is in contrast to TLR-activating immunostimulatory RNAs that are bound in the endosome. For the delivery of ppp-siRNAs into the cytosol, transfection reagents are required, whereas TLR-stimulating RNAs are applied without transfection reagents. Apart from the triphosphate moiety at the 5′ end, ppp-siRNAs share the same structure as siRNAs. Therefore, transfection reagents that are effective for siRNA delivery are likely to be effective in the application of bifunctional RNAs. Numerous different reagents, mainly based on liposomal formulations, are commercially available. Their efficacy strongly depends on the cell type. Therefore, the transfection conditions have to be established. A general starting point for ppp-siRNAs is liposomal transfection and a RNA concentration of 1 μg/ml. The typical readouts to measure efficacy are stated in the sections below.

Because of the low bioavailability of RNAs, the systemic delivery of ppp-siRNAs (and siRNAs) is much more demanding. Here, polyethylenimine derivatives are often used [11–13]. Significant improvements were achieved by chemical modification of the RNA and by complexation with special liposomal particles and other nanoparticles; however, many of them are not commercially available (for a review see ref. [14]).

3.2.2 Determination of the Efficacy of Bifunctional RNAs

Determination of RNA Interference Efficacy

Active RNA interference results in cleavage of the targeted mRNA close to the binding site with subsequent degradation of the cleaved mRNA. Efficacy of the gene silencing property of bifunctional RNAs can be analyzed on the mRNA and protein level. The advantage of measuring on mRNA level, e.g., by quantitative RT-PCR, is that mRNA degradation during RNAi is independent of mRNA and protein stability. RNA degradation depends on the RNAi apparatus of a given cell type, but not on the targeted gene. If the RNAi kinetics for that cell type is known, the efficacies of siRNAs can be compared and estimated for new genes. RNAi kinetics depend on the cell type, but for many cell types 24 h after transfection is a good starting point for mRNA quantification.

In almost every case, reduction of protein levels is required to initiate the corresponding biologic consequences. Therefore, measuring on protein level, e.g., by immunoblotting or by ELISA, is essential.

However, protein reduction does depend not only on RNAi but also on the specific half life of the protein. This strongly varies among proteins. As a starting point, we control for protein reduction 48 h after siRNA treatment. Additionally, performing time course experiments can be necessary to identify the earliest time point of significant protein reduction. In further steps, the functions associated with the targeted gene can be evaluated.

RIG-I-Mediated Effects

RIG-I promotes induction of interferons, proinflammatory cytokines, and proapoptotic proteins. Various methods can be used for analyzing this. Analysis of induction of interferons or proinflammatory cytokines is mostly quantified in the cell culture supernatant by ELISA for example for IFN-α, IFN-β, or IP-10 (CXCL10). Different ELISA kits are commercially available. Alternatively, IFN levels can be analyzed on the mRNA level by quantitative RT-PCR. As a representative interferon, IFN-β has mostly been used. mRNA quantification is a sensitive method and it allows for parallel quantification of other molecules. For example, the two key functions of bifunctional siRNAs—immunostimulation and gene inhibition—can be analyzed simultaneously (*see* **Note 1**).

RIG-I induces the proapoptotic molecules Noxa and Puma. Noxa was identified to be critical for RIG-I-dependent apoptosis in melanoma cells [4]. Both molecules are induced on RNA level and qRT-PCR or immunoblotting are appropriate methods to measure proapoptotic activity. Depending on the apoptotic activity in the given cell type, quantification should be done at the beginning of cell death induction, e.g., 17-48 h after transfection (*see* **Note 2**).

3.2.3 Measuring Cell Death in Tumor Cells

Several methods exist that help to characterize the type of cell death.

Measurement of Cell Viability

The amount of viable cells can be determined by quantifying their metabolic activity, assuming that vital cells have a typical metabolism (*see* **Note 3**). Among others, fluorescence-based assays based on the cell permeable dye resazurin are widely used. Viable cells with intact metabolism are able to reduce resazurin to fluorescent resorufin. The fluorescence signal is proportional to the amount of viable cells in the sample. The conversion of resazurin can be monitored visually as the color changes from blue to pink.

Materials:

- Resazurin-based dye (commercial kits).
- Fluorescence reader with excitation 530–570 nm and emission 580–620 nm filters (*see* **Note 4**).
- 96-Well tissue culture plates compatible with fluorometer (clear or solid bottom).

Method:

1. The assay can directly be performed in the culture plates used for transfection of the bifunctional siRNAs. Briefly, the dye is added to the medium of transfected cells. Dependent on the metabolic activity, the plate is incubated for approximately 15–60 min at 37 °C.
2. Controls:
   - No-cell control (background medium fluorescence, blank value): Add resazurin-based dye to sterile medium (without cells) in the same ratio as in treated samples.
   - Untreated control cells: Set up wells with untreated cells serving as a vehicle control and add the same amount of reagent as to samples.
3. Regularly check for a change in color as compared to controls. Upon color change to purple-pink, stop the reaction by transferring 100 μl of supernatant in a well of a 96-well plate for analysis and measure fluorescence (*see* **Note 5**).
4. The fluorescence values can be normalized and displayed as percentage compared to the positive control (set to 100 %).

Measurement of Apoptosis

Determination of Annexin V-Binding. Alterations in the cell membrane are characteristic for apoptotic cells. During apoptosis, phosphatidylserine, normally located at the inner layer of the cell membrane, translocates to the outer membrane layer. This can be measured by Annexin V (AN), a non-cell-permeable protein that binds phosphatidylserine. The measurement can be completed by using propidium iodide simultaneously. Propidium iodide stains DNA and is not cell permeable. Dead or damaged cells are characterized by a leaky membrane, allowing propidium iodide to enter cells. This enables distinguishing between viable cells (AN-negative/PI-negative), cells currently undergoing apoptosis (AN-positive/PI-negative), and dead/damaged cells (AN-positive/PI-positive). Typically, quantification is carried out via flow cytometry (FACS). *See* also Fig. 3.

The measurement of Annexin V binding complements cell viability assays due to two reasons: (1) in cell viability assays, the cause of cell number reduction is not determined, and (2) in cell viability assays the overall reduction of viable cells during the timeframe of the experiment is measured. This allows us to determine the overall cytotoxic effect of bifunctional RNA as a result of cell death. In contrast, in Annexin V assays the percentage of apoptotic and dead cells at a given time point is determined.

Materials:

- Fluorescent-labeled Annexin V (commercial); e.g., fluorescein isothiocyanate (FITC).

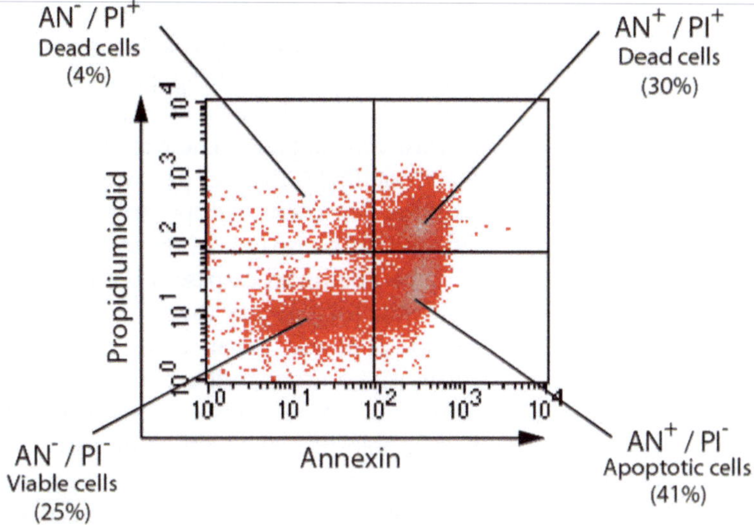

Fig. 3 Example of a dot plot of a FACS measurement after staining with Annexin V and propidium iodide. AN^{+}/PI^{-} cells can be interpreted as being apoptotic. AN+/PI+cells represent dead cells. These populations likely died via apoptosis, but this staining pattern does not provide evidence for apoptosis. Due to their leaky membrane these cells stain also positive for Annexin V. For total quantification of dead and apoptotic cells, the sum of AN^{+}/PI^{-}, AN^{+}/PI^{+}, and AN^{-}/PI^{+} populations may be formed. *AN* Annexin V, *PI* Propidium iodide

- 6× Propidium iodide solution (6 μg/ml) in PBS.
- Staining buffer (10 mM HEPES, 140 mM NaCl, 5 mM $CaCl_2$): For 50 ml of buffer: 500 μl 1 M HEPES, 1.4 ml 5 M NaCl, 125 μl 2 M $CaCl_2$, 48 ml double-distilled H_2O, adjust pH to 7.4 with NaOH.

Method:

1. To collect detached cells, collect supernatant medium of each well and pellet cells by centrifugation (*see* **Note 6**).
2. To collect adherent cells detach and transfer cells to tubes containing the pelleted supernatant cells obtained in **step 1**.
3. Wash cell pellets 3 times with PBS to ensure that the cell viability dye is completely removed.
4. Prepare 1× Annexin V solution according to the manufacturers' protocol and resuspend pellet in 50 μl of 1× Annexin V solution (*see* **Notes** 7 and **8**).
5. Incubate for 15 min at room temperature in a dark place.
6. Prepare 1× propidium iodide solution: 10 μl propidium iodide + 50 μl Staining buffer (final concentration of 1 μg/ml) per sample and add to Annexin V-stained cells.
7. Analyze via flow cytometry immediately after adding propidium iodide.

Determination of Caspase Activation. Caspases are proteolytic enzymes that play a key role in apoptosis and caspase activation is a key feature to determine apoptosis. Caspases are activated by proteolytic cleavage in the form of a cascade where activated caspases cleave downstream caspases. There are two mayor apoptotic pathways, the mitochondrial (intrinsic) pathway and the death receptor (extrinsic) pathway. The respective apoptosis pathway leads to activation of specific initiator caspases. Caspase 9 is typical for the mitochondrial apoptosis pathway, caspase 8 is critical for the death receptor pathway. Initiator caspases then activate common effector caspases like caspase 3, 6, and 7 that carry out apoptosis. Caspase activation can be measured using antibodies that detect the proform as well as the active cleaved subunits (*see* **Note 9**). These can be visualized by immunoblotting due to the different size of proform and subunits. An alternative way is the use of fluorescent substrates that are specific for a certain caspase. Activation of RIG-I leads to mitochondrial apoptosis leading to activation of caspase 9 and effector caspases [4]. Therefore, measuring activation of caspase 9 and caspase 3 as an effector caspase is suitable (*see* **Notes 10** and **11**).

4 Notes

1. The optimal time point of measurement is critical and depends on the cell type. Increased IFN-β mRNA levels can be measured earlier than the effects of RNAi, usually within a time frame of 6 to 24 h. The apoptotic effects have to be taken into consideration when choosing longer time periods as cell death can result in RNA of insufficient quality. As a starting point, quantification 17 h after transfection may be suitable for many cell types.
2. Proapoptotic signaling via RIG-I seems to be independent of p53. Both molecules are well known as downstream targets of the tumor suppressor p53. However, p53 was dispensable for RIG-I-mediated apoptosis and induction of Noxa was independent of p53.
3. Cells currently undergoing the process of apoptosis with strong inductors of apoptosis (e.g., staurosporine) can display an increased metabolic activity compared to healthy cells.
4. As an alternative to fluorescence, absorbance can be used for quantification; however, the sensitivity is less and more cells have to be used.
5. After measurement, the cells can be re-used for other analyses, e.g., for apoptosis induction by staining with Annexin V. For this, the cells have to be thoroughly washed (3 times with PBS) to remove the fluorescent dye.
6. It is important for adherent cell types to include detached cells in the supernatant, as they represent dead cells. If the supernatant is

discarded, the fraction of apoptotic and dead cells may be underestimated.

7. Annexin V and propidium iodide are light sensitive and have to be stored in a dark place.
8. In many cases it is possible to reduce the incubation volume of the Annexin V-containing buffer, which reduces the amount of Annexin V required for the assay.
9. For caspase immunoblotting, antibodies that solely detect the proforms are not suitable for apoptosis measurement. Antibodies specific for cleaved subunits can be used, but they need quantification of a house-keeping gene as a loading control. Antibodies that detect caspase proforms as well as their subunits are most convenient. They do not require quantification of a house-keeping gene, because detection of the proform allows estimating gel loading.
10. At late stages of apoptosis or upon strong apoptosis inducers most caspases become activated due to a general high proteolytic activity. In this case, it is not possible to distinguish between apoptosis pathways, because most initiator caspases are activated. An alternative way in this situation is it to inhibit initiator caspases by RNAi or caspase inhibitors and to test which initiator caspase is responsible for apoptosis.
11. As an alternative to immunoblotting, caspase inhibitors are available. This includes Z-VAD-FMK, an inhibitor of all caspases that is able to rescue cells from caspase-dependent apoptosis, and specific inhibitors, e.g., for initiator caspases like Z-IETD-FMK (caspase 8) and Z-LEHD-FMK (caspase 9).

Acknowledgement

This work was supported by the German Cancer Aid (grant 107805) and by the Melanoma Research Network (German Cancer Aid) to RB and by the German Research Foundation (DFG) grant GK 1202 to FM and RB.

References

1. Akira S, Uematsu S, Takeuchi O (2006) Pathogen recognition and innate immunity. Cell 124:783–801
2. Fire A, Xu S, Montgomery MK, Kostas SA, Driver SE, Mello CC (1998) Potent and specific genetic interference by double-stranded RNA in Caenorhabditis elegans. Nature 391: 806–811
3. Elbashir SM, Harborth J, Lendeckel W, Yalcin A, Weber K, Tuschl T (2001) Duplexes of 21-nucleotide RNAs mediate RNA interference in cultured mammalian cells. Nature 411:494–498
4. Besch R, Poeck H, Hohenauer T, Senft D, Hacker G, Berking C et al (2009) Proapoptotic signaling induced by RIG-I and MDA-5 results in type I interferon-independent apoptosis in

human melanoma cells. J Clin Invest 119: 2399–2411

5. Glas M, Coch C, Trageser D, Dassler J, Simon M, Koch P et al (2013) Targeting the cytosolic innate immune receptors RIG-I and MDA5 effectively counteracts cancer cell heterogeneity in glioblastoma. Stem Cells 31(6): 1064–1074
6. Hornung V, Ellegast J, Kim S, Brzozka K, Jung A, Kato H et al (2006) 5'-Triphosphate RNA is the ligand for RIG-I. Science 314:994–997
7. Ellermeier J, Wei J, Duewell P, Hoves S, Stieg MR, Adunka T et al (2013) Therapeutic efficacy of bifunctional siRNA combining TGF-beta1 silencing with RIG-I activation in pancreatic cancer. Cancer Res 73: 1709–1720
8. Reynolds A, Leake D, Boese Q, Scaringe S, Marshall WS, Khvorova A (2004) Rational siRNA design for RNA interference. Nat Biotechnol 22:326–330
9. Ui-Tei K, Naito Y, Takahashi F, Haraguchi T, Ohki-Hamazaki H, Juni A et al (2004) Guidelines for the selection of highly effective siRNA sequences for mammalian and chick RNA interference. Nucleic Acids Res 32:936–948
10. Yuan B, Latek R, Hossbach M, Tuschl T, Lewitter F (2004) siRNA selection server: an automated siRNA oligonucleotide prediction server. Nucleic Acids Res 32:W130–W134
11. Urban-Klein B, Werth S, Abuharbeid S, Czubayko F, Aigner A (2005) RNAi-mediated gene-targeting through systemic application of polyethylenimine (PEI)-complexed siRNA in vivo. Gene Ther 12:461–466
12. Morrissey DV, Lockridge JA, Shaw L, Blanchard K, Jensen K, Breen W et al (2005) Potent and persistent in vivo anti-HBV activity of chemically modified siRNAs. Nat Biotechnol 23:1002–1007
13. Poeck H, Besch R, Maihoefer C, Renn M, Tormo D, Morskaya SS et al (2008) 5'-Triphosphate-siRNA: turning gene silencing and Rig-I activation against melanoma. Nat Med 14:1256–1263
14. Whitehead KA, Langer R, Anderson DG (2009) Knocking down barriers: advances in siRNA delivery. Nat Rev Drug Discov 8:129–138

Index

Hans-Joachim Anders and Adriana Migliorini (eds.), *Innate DNA and RNA Recognition*, Methods in Molecular Biology, vol. 1169, DOI 10.1007/978-1-4939-0882-0, © Springer Science+Business Media New York 2014

S

T

U

V

X

Z

MIX
Papier aus verantwortungsvollen Quellen
Paper from responsible sources
FSC® C105338

If you have any concerns about our products,
you can contact us on
ProductSafety@springernature.com

In case Publisher is established outside the EU,
the EU authorized representative is:
Springer Nature Customer Service Center GmbH
Europaplatz 3, 69115 Heidelberg, Germany

Printed by Libri Plureos GmbH
in Hamburg, Germany